SlipString Drive

SlipString Drive

String Theory, Gravity, and "Faster Than Light" Travel

Andrew L. Bender

iUniverse, Inc.
New York Lincoln Shanghai

SlipString Drive

String Theory, Gravity, and "Faster Than Light" Travel

iUniverse books may be ordered through booksellers or by contacting:

iUniverse
2021 Pine Lake Road, Suite 100
Lincoln, NE 68512
www.iuniverse.com
1-800-Authors (1-800-288-4677)

Edited by Dr. Lee Bender, David Bender, Dr. Carol Schilling, and Norma Torney

ISBN: 978-0-595-40822-1 (pbk)

ISBN: 978-0-595-85206-2 (cloth)

ISBN: 978-0-595-85185-0 (ebk)

Printed in the United States of America

For my family
To my parents, brother, and grandparents
who have inspired me and without whom
this book would not be possible.

Contents

Acknowledgments xi

Preface xiii

Chapter 1: String Theory Preliminaries 1

Chapter 2: M-Theory, where "M" Is for "Membrane" (also Mega, Magic, Matrix, or Mother-of-All…) 14

Chapter 3: The Motions of Isolated Volumes of Spacetime or "SlipString Drive" 17

Chapter 4: UFOs as Proof-of-Concept? (No, I am not trying to prove they exist!) 35

Chapter 5: Exploration Opportunities Using SlipString Drive 43

Chapter 6: Gravity Waves and How to Manipulate Them 51

Chapter 7: SETI (Search for Extraterrestrial Intelligence) and How to Fix It 59

Chapter 8: How to Complete a "Theory of Everything" by Modifying M-theory with a Membrane Theory of Gravity 66

Chapter 9: Life, the Universe and Everything, or: How the Universe Will End…But Must We End With It? 94

Appendix Preface 99

Appendix A Hypothesis for a Membrane Theory of Gravity 101

Notes 129

Appendix 133
Bibliography 135

The important thing is not to stop questioning. Curiosity has its own reason for existing.

—Albert Einstein

Acknowledgments

Thanks to Dr. Lee Bender, David Bender, and Dr. Carol Schilling for all their work of editing and revising the book while in progress and for helping to refine my work.
Thanks to Norma Torney for her final proof and edit of the book.
Thanks to Samuel T. Swansen, PC, for his faith in my work and for pushing to get it published.
Thanks to Theodore C. Nason Ph.D. for helping to refine my theories and for a critical analysis of them.
Thanks to NASA, ESA, R. Sahai and J. Trauger (Jet Propulsion Laboratory) and the WFPC2 Science Team for the image of the Boomerang Nebula used on the front cover.
Thanks to the WMAP Science Team for the WMAP cosmic background radiation image in chapter 8 and in Appendix.
Thanks to Dr. Debra Elmegreen for her critique of the "Membrane Theory of Gravity."
And thanks to Drs. Fred Chromey, Debra Elmegreen, James Lombardi, and Sophie Yancopoulos among other faculty and staff, for their help and inspiration while at Vassar College and for their continuing friendship.

Preface

Ever since I was six years old, I have been fascinated by how the universe works. I was glued to the TV series *Cosmos* with astronomer Dr. Carl Sagan. When I was about nine years old, for a class project I wrote "Andrew's Galactic Believe-It-or-Not" with the latest astronomical oddities and interesting objects that had been discovered by the late 1970s to early 1980s. I was especially intrigued by pulsars, quasars, black holes, and the vastness of the universe. I was always asking myself big questions about how the universe was really put together.

Recently, I have been developing new theories about the universe and how to take advantage of these assumed properties to do things like escaping from the end of the universe and possibly to travel "faster than light." I speculated in 1999 that in order to avoid ending with the universe we might be able to create a bubble of strings (which compose our universe) around a ship and leave this universe for a younger, healthier one. In 2003, on TechTV's *Big Thinkers* series, I was encouraged to learn that Dr. Michio Kaku postulated the same idea about leaving our universe within a bubble of strings.

After much thought on the beginning of the universe, I also had a hunch that the universe had begun through a collision of massive proportions rather than the much more popular belief—the big bang theory—that the universe started from a single infinitesimal point that inflated into our current universe. I had problems with the traditional big bang theory because of its lack of explanation for how and why it occurred and thought that only a massive collision could produce such an event. This idea was later supported in a *Science* magazine article in 2001 that postulated the idea that string theory (M-theory, to be specific) could explain the big bang through the collision of huge membranes (or "branes" for short), which float around in the higher spatial dimensions proposed in M-theory and could cause a "big splash" when the membranes collided, creating a new universe.[1] According to M-theory, these collisions were not necessarily uncommon either. Although I could not yet see the mechanism by which the collisions occurred (without a fuller understanding of string/M-theory), I appreciated the possibilities. The idea that an entire universe could sprout from an almost one-dimensional point for no particularly good reason seemed unlikely, but a collision of membranes of massive sizes and energies seemed much more plausible. Plus, this theory avoids some of those nasty infinites that keep popping up in cosmological theories and are usually a sure sign of difficulties with a theory.

At least a half dozen times, theories I postulated years earlier were later promoted by other scientists, and most have been incorporated into current scientific thinking. So I finally decided to sit down (well, I am pretty much always sitting down) and get to work on publishing my latest theory of travel that could be effectively "faster than light." I spent several thousand hours mentally modeling the universe and running "simulations" on those models. The more I understood string theory and the more my theories progressed on how it might be possible to create a back door around the limitations of the speed of light, the more "eureka" moments I had about the creation of our universe. I started putting the pieces of the puzzle together and eventually came across an elegant answer to what gravity might be. Once I had that piece, the entire universe started to make sense, and now every new astronomical finding that baffles cosmologists, such as dark matter, dark energy, and the universe's increasing and changing rate of expansion, and many other discoveries, all appear to fit squarely into this new theory. If correct, these modifications of M-theory should finally turn it into a true "Theory of Everything."

You may be wondering why someone with a bachelor of arts in astronomy is working on such matters. Well, it is more a matter of time for me. I suffered a spinal cord injury as a teenager in 1988, as a result of a skiing injury. Not only am I a quadriplegic (although I did regain most of my upper body strength and function, some hand function, and near normal sensation everywhere), but I also developed chronic lower back pain. While all these factors have slowed down my work, I did have time to read and study but most of all to think.

In 1996 after six years of taking courses at a local community college (pain permitting), I was finally able to get my back into good enough shape that I could transfer to Vassar College, where I was the college's first student in a wheelchair to live on campus. I received my BA in astronomy from Vassar in 2000 and have continued my studies in string and M-theories ever since.

So here I am in my early thirties with my education delayed but itching to share my knowledge. I must get these theories to the public before another decade flies by, because I am not getting any younger, and I have not and will not come up with a practical theory for time travel!

I am also developing a video game based on the book that will span the life of our universe and beyond (in three separate and highly entertaining stages), so that readers can experience the travel described in the book for themselves. In addition, I would like to further develop my theories in an academic setting.

CHAPTER 1

String Theory Preliminaries

I grew up in an Einsteinian universe. I could see spacetime being curved by gravity but was always wondering about the underlying nature of the universe. What exactly were matter and radiation, and how did they travel through a vacuum? I needed a more satisfactory explanation for what gravity was exactly and why light supposedly has a dual nature, sometimes acting as a particle and other times as a wave. Theorists struggled to explain this puzzling duality. Also relativity (Einstein's theories of gravity and light which apply on cosmic scales as matter approaches the speed of light and where massive gravitational forces are concerned) and quantum mechanics (trying to describe what happens on a small scale among atoms, electrons, quarks, photons, and even "empty" space itself) are so diametrically opposed that they appeared not to be describing the same universe. Quantum mechanics is not as elegant as relativity, is often quite counterintuitive, and cannot currently be integrated into Einstein's theories of relativity (it is similar to trying to duct tape a rabbit to a turtle and calling the outcome a single animal). Quantum mechanics has so much unstable quantum noise on very tiny levels (scales less than the Planck length[1] of 10^{-33}cm or 10^{-35}m) that the universe would be a relatively unstable place if it operated completely by this theory. Thus, very few physicists believe that quantum mechanics could possibly be a final theory, or Theory of Everything (TOE), to describe the universe, even though it has had a good deal of success in its predictive abilities of the statistics of the universe. For example, it predicts quite accurately the approximate number of photons that, when sent through a particular diffraction grid, will strike a given area on the other side of the grid. However, it cannot give precise predictions about what individual photons in the group will do.

String theory has been under development in one form or another since the 1970s, but it was not really taken seriously by many mainstream scientists until the 1990s when many of its kinks were worked out. String theory looked like the only feasible game in town for a TOE now that quantum theory was shown to have several weaknesses the deeper you looked into it. It made so much more sense to me than trying to unite two seemingly incompatible theories in order to attempt to explain one universe, a universe in which you cannot always tell what is a particle and what is a wave. String theory eliminates all the point particles. Everything is basically a wave, or string vibration, in this theory.

Previously, physics described matter, energy, and most forces as the effects of point particles. Light, for example, was described as photons which were individual points or packets of light. The confusing part was that light interferes with itself like water and other waves do, generating interference patterns of dark and light bands where the waves destructively and constructively interfere with each other, just like a wave does. Electrons and quarks were thought of as individual points of matter, and even forces of nature were thought to be transferred by subatomic particles such as gluons which hold quarks together inside protons and neutrons in the nucleus of an atom. With string theory, now everything (all matter, energy, and forces) can be described as a wave function, or as the side effects of the vibrations of strings of energy which vibrate in ten different dimensions (or in eleven dimensions with M-theory: more on that later).

Each string vibrates in the three spatial dimensions we know well plus a time dimension, in addition to six (or seven) more spatial dimensions. These exact numbers of dimensions were necessary mathematically in order to model all the forces of nature within the theory. Only with at least six extra spatial dimensions could the vibrations of strings take on all the characteristics of matter, light, and all the forces in nature. If there were any fewer dimensions, the strings would not have enough room to vibrate and therefore could not generate all the different forces and types of matter that we observe in our universe.

A precursor to string theory attempted to unify all the forces in physics simply by adding more spatial dimensions to the three familiar ones. While that theory was incomplete without strings, it gives you a good idea of how extra dimensions help to describe our universe. In April 1919, long before the development of string theory (and even before quantum theory), many physicists took seriously Theodor Kaluza's theory that light is a vibration of the "fourth spatial dimension" because it finally unified gravity and light within a single framework.[2] The theory was mathematically elegant, and it was not easily dismissed by critics at the time. However, in the 1930s, most physicists moved away from this concept and began to research the recently developed quantum theory, which predicted quantum-mechanical effects (characteristics of the very small, such as atoms, electrons, and quarks) with highly accurate probabilities.[3] However, quantum mechanics only gives probabilities, and it is exceedingly difficult to "pin down" a solid number using this theory which makes it quite different from the physics that came before it.

So why does a multidimensional string theory make more sense than previous theories? First of all, it explains the large and small with equal clarity because there is a mathematical limit as to how small a string can be (a string's size is the Planck length mentioned earlier of 10^{-33} cm, or about 100 billion billion times smaller than a proton!).[4] Many have wondered what kind of medium light and gravity travel through, and string theory gets much closer to a satisfactory explanation. The universe is made up one-dimensional loops of string vibrating in ten dimensions (or in M-theory two-dimensional strings, which I will describe in the next chapter) whose three spatial and one temporal dimension have expanded to the macroscopic scale (we can see or at least sense those dimensions around us on an everyday level). All of the extra spatial dimensions that strings vibrate in are so tiny that humans are much too big to perceive or even detect them, and so we have not previously concerned ourselves with them. However, these extra-dimensional vibrations and the resulting shapes of the strings in those higher dimensions account for all of the observed forces and types of matter in the universe.[5]

To understand how there can be so many extra tiny dimensions that we cannot see, imagine looking at a telephone line hanging between two poles far off in the distance. The line strung between two telephone poles appears to be a one-dimensional string on which an insect could only travel forward or backward along its length. However, as we move in closer to the telephone line, we realize that its surface actually has two dimensions (we are not counting the inside of the cable, just its outer surface). Looking closely at the cable, we now see that an insect can actually move in two dimensions. They can walk both backward and forward along the length of the line, as well as being able to walk around the circumference of the cable, to the left or to the right. In this manner, these extra tiny surfaces are how additional dimensions (if wrapped up into very tiny spaces like the Planck length) can exist without humans being aware of them. Because strings are so much smaller than the protons or even the spaces between the quarks inside an atom, they can curl up into a multitude of dimensions and directions that we have no way of perceiving with our senses or current technology. However, mathematics has begun to reveal their possible nature and describes what they may look like.

These vibrating strings of energy are much too small to detect with current technology because they are much smaller than the effects of the forces that they generate. We observe an electron to be a certain size because of its forces, but we cannot look inside the electron to see what it is made of or what it looks like as we do not have anything smaller than an electron to shoot at it to reveal its shape or size as we do with electron microscopes. An electron microscope produces images by shooting electrons at a target, and because the electrons are so small, they bounce off the atoms in the target, revealing the details of an object all the way down to resolutions as small as a single atom.

The amount of energy it would take in a particle accelerator to smash our way down to a single string is so high that there is no way (at least in the next few hundred years) to generate that kind of energy in a collision. However, there may be other experimental or even astronomical ways to either infer the existence of strings or to see their imprint on the expanding cosmos from its original inflation as I will describe in chapter 8. One experimental way to prove M-theory (an expansion of string theory) is by a "disappearing graviton," as I will explain in chapter 2.

Long ago the Pythagoreans thought there was something musical about the heavens. They realized that energy was quantized (as in quantum) into discrete increments, like notes on a scale, and they thought that the universe was playing out like a symphony. It turns out they were probably right! As with the vibrating strings on a piano, we can only create stable vibrations on each harmonic of a string, as with the strings that are pinned down on each side in fig. 1.1). If one string strikes another string gently, it will not have enough energy to raise the string's frequency to the next stable harmonic resonance, like changing from the top string's frequency to the second string's frequency (see fig. 1.1). Because there is not enough energy to raise the string's frequency to the next energy level, both strings would stay at their original frequency (or note) and simply bounce off each other at the same speed at which they originally collided.

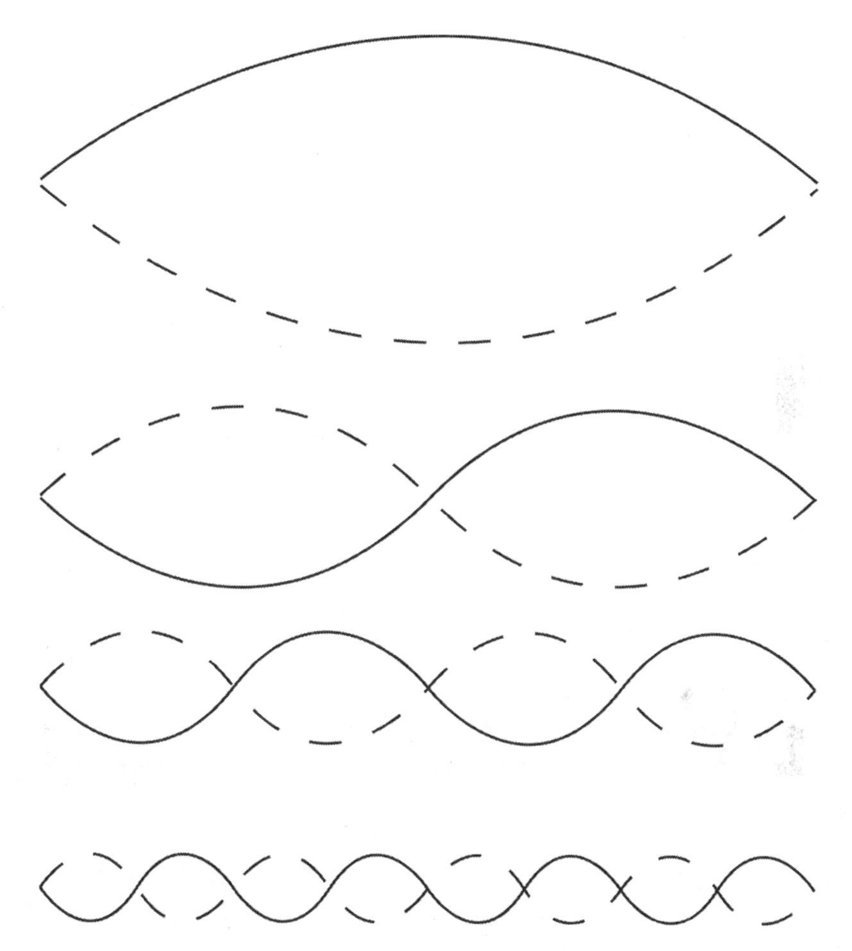

Fig. 1.1. *Harmonic frequencies of a vibrating string. These are the first, second, fourth, and eighth harmonic frequencies of a one-dimensional string. This is similar to shaking a jump rope up and down in rhythm with a partner. The harder and faster both of you shake the ends of the rope, the higher the frequency of the waves you create.*

However, if one string is traveling fast enough when it collides with the second string, the second string will increase its frequency by one period or more (depending on the speed of the collision), and the momentum of the first string will be reduced by the amount of energy that was transferred to the second string. Because of this, there are discrete energy levels of strings, and each higher frequency (harmonic note) of a string produces the characteristics of a different and more massive form of matter. All matter and energy are made up of strings that simply vibrate at different frequencies and therefore create different shapes as they vibrate in their ten (or eleven) different dimensions.

Alternatively, instead of strings that are pinned down on either end, consider strings that vibrate as loops (see fig. 1.2). This is a simplistic one-dimensional representation, but it demonstrates the nature of these loops. (There are also strange-looking six-dimensional mathematical constructs that have been created on computers to give us a better idea of what actual strings might look like which you can find on the Web.) In 1974, Scherk and Schwartz determined that the gravitational force was inversely proportional to the tension of each string.[6] Because gravity is such a weak force on each individual string (and over long distances), the tension of each string was calculated to be 10^{39} tons, making these strings extremely stiff. This is the so-called Planck tension. This severe tension means that each string is quite small (10^{-33} cm), and that it takes a significant amount of energy to start a string vibrating to create matter and energy.[7]

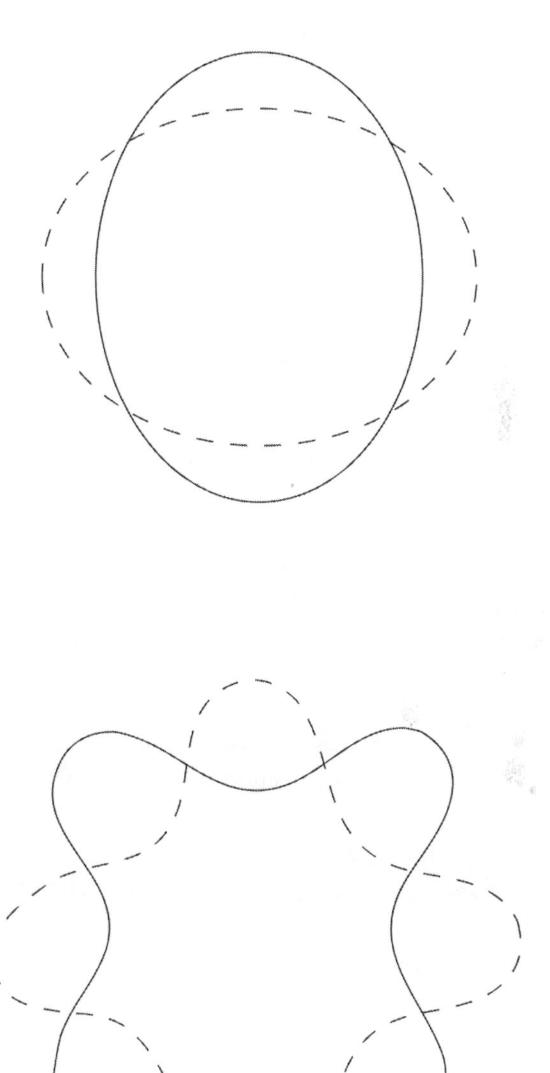

Fig. 1.2. *Harmonic frequencies of vibrating loops of string.*

Brian Greene, in his book *The Elegant Universe,* describes string theory in a very understandable manner. I recommend reading the book or watching the PBS *Nova* TV special of *The Elegant Universe.* Greene thoroughly explains the evolution of a TOE—from Relativity to Quantum Mechanics to string theory—and provides helpful visual explanations. Dr. Michio Kaku's book, *Hyperspace,* is another excellent explanation of string theory, its roots, how it was developed, and Dr. Kaku's role in helping to expand and refine it. I highly recommend both of these books if you are interested in more details about string theory and its evolution.

Now, what about gravity? How is it transmitted? What is it, really? According to string theory, gravity is a slight curvature of spacetime due to the vibrations of strings that act as matter. Because it bends space, massive bodies fall toward one another, similar to an apple falling from a tree. It is the weakest of all forces, which can be easily demonstrated by jumping up in the air. Your body just overcame the entire gravity of every atom comprising planet Earth pulling down on you. When your feet hit the ground on your way down, the tiny electrons that circle the atoms in your feet were stopped instantaneously by the electrons orbiting the atoms in the ground, and there was nothing gentle about your stop, unlike your rise and fall. This abrupt and forceful landing demonstrates the strength of the electromagnetic force, as you do not sink into the ground but are strongly repelled by it. The few electrons in your feet (compared to the enormous number of atoms within planet Earth) strike the ground's electrons (as opposed to your relatively slow rise and fall from the force of gravity). M-theory has an explanation for this weakness of gravity as described in chapter 2, and I will propose a new hypothesis for gravity itself within an M-theory framework in chapter 8 that may finally explain the mystery of gravity once and for all, creating a true M-theory TOE.

Further, light waves and all radiation are just the pluck of a string. Such radiation travels at the speed of light ($C = 3.0 \times 10^8$ meters/second, or about 186,000 miles per second) through the string medium, which is not unlike a very thin liquid. Several of Einstein's relativistic equations are actually quite similar to the equations for fluid dynamics, such as the relativistic effect of frame dragging (which I will expand on later). It is as if the earth is spinning in and passing through a liquid, which is actually just the string medium of which our universe is made. These strings make up the fabric of space and explain all relativistic effects, such as time slowing down and mass increasing as matter attempts to approach the speed of light. There is a detailed description of these effects and how they could be created in chapter 3, although physicists have not yet sufficiently addressed the question of why these effects occur.

As with sound waves that travel at the speed of sound (742mph @ 0°celsius, 790mph @ 35°celsius)[8] along pressure waves through our gas atmosphere (its medium), the speed of light is simply the rate at which strings without mass (light waves) propagate through the string medium of our universe. The strings that make up light waves have no mass and are pure energy because the shape of their string vibrations only intersects two of our three spatial dimensions and do not vibrate in the third dimension of gravity. Photons also vibrate in one higher dimension that we cannot sense, and they cannot have any mass because they are absolutely flat (two-dimensional) in our universe. You can observe this effect by looking across the beam of a flashlight or laser beam. You cannot actually see any light from the beam if you are standing directly perpendicular to it. You can only see light from the beam in this position if there is sufficient dust or fog in the air along the beam's path to bounce photons off and in your direction so you can observe them. Otherwise, all the photons just travel perpendicular to you until they hit a wall or something else directly in front of the beam, and then they scatter in all directions after hitting the wall, illuminating the room.

According to Kaku, if you were able to strike the string that makes up a quark (as far as we can tell one of the smallest forms of matter in the universe that makes up protons and neutrons) with enough energy, you would get a different form of matter: an electron. If you were able to strike the string that made up this electron with the proper amount of energy, you would get a photon (a wave of light and pure energy). All three types of matter and energy are just the same fundamental string vibrating at three different frequencies. These different frequencies cause the string to change shape and vibrate more or less in different spatial dimensions, giving each configuration different characteristics (masses, charges, and strengths of each characteristic). This is similar to how the strings in fig. 1.1 vibrate at different frequencies, but they would vibrate at different angles (dimensions) as well. Because strings vibrate in so many different dimensions, their vibrations give unique properties to each different shape that the string takes on as its frequency increases, which gives the quarks, electrons, and photons (as well as many other short-lived exotic particles) such different properties.

These three different configurations of strings make up all the different types of normal matter and energy in the universe. The quark is the most massive and slowest moving configuration of normal matter because it warps spacetime the most. (The more massive something is, the more gravity it has, and the more that it warps space and time.) The electron has much less mass and can travel at or near the speed of light, depending on the medium it is traveling through). The photon has no mass whatsoever and zips along at the universe's maximum speed of C (again, C = the speed of light = 3.0×10^8 m/s). According to Greene, the more energy a string has (the higher its frequency), the more massive the particle the string emulates.[9] Photons are also more energetic the higher their frequency. However, because their vibrations only intersect two of our three spatial dimensions, they have no mass.

The most common types of quarks are the up and down quarks with electrical charges of +2/3 and −1/3, respectively. These quarks join together in groups of three with either two-up and one-down quark for a charge of +1 (a proton) or with two-down and one-up quark for a charge of 0 (a neutron). All atoms are simply different combinations of protons, neutrons, and electrons, which are really just various configurations of vibrating strings in different combinations and numbers. For example, hydrogen is the simplest atom with a single proton in the nucleus and only one electron orbiting it. This configuration makes the atom stable, and it carries no charge because it has the same number of protons as it has electrons (which have a charge of −1). Helium is the next heaviest element with two protons and two neutrons in the nucleus and two electrons orbiting the nucleus. Each heavier atom simply has more protons and neutrons in the nucleus and the same number of electrons as it has protons, so it has a neutral charge of 0.

It is possible to remove electrons from an atom, creating an ion and giving the atom a positive charge, or to add electrons, giving the atom a negative charge. If you keep fusing these atoms together (as is done with the massive gravity in the heart of a star and again when a star goes supernova and explodes), you can create all the forms of matter (atoms from hydrogen to uranium) in the universe. Because of their different characteristics, these atoms can interact and join together, forming complex compounds with even more complex characteristics. Such complex molecules will interact with each other and, with the assistance of lightning, solar wind, deep sea thermal vents, and other energetic processes, become organic compounds such as amino acids, which are the building blocks of life. When amino acids are compressed together if a comet or asteroid strikes the earth, they become much more complex organic molecules: peptides. Through trillions of random interactions and with energy given to these compounds through thermal vents at the bottom of the ocean under intense pressure and with chemicals spewed from these vents (or other energetic processes), eventually self-replicating organic compounds form. Over time, these compounds combine, compete, and evolve to contain RNA and then DNA, and they will eventually evolve to form life as we know it. Thus, an extremely complex system has evolved from very, very simple beginnings.

There are other types of quarks called the charm, strange, bottom, and top quarks. However, they only exist for brief moments in time in atom smashers due to the massive amounts of energy necessary to create them. In nature they are created in high-energy cosmic collisions, supernovae, and other high energy interactions before they break down into more stable configurations of strings—matter and energy. (For the purpose of this discussion, we do not need to go into detail of these exotic and short-lived forms of matter.)

Additionally, because of the nature of strings and their ability to vibrate in opposite waveforms that cancel each other out when combined, there is "supersymmetry." This means that for every kind of particle (actually a string vibration) there is an opposite form of the vibration. For example, the electron has the oppositely charged positron as its antiparticle (or anti-resonance), and each quark has its anti-quark partner, and so on. These types of opposite string configurations do not last long in nature before they annihilate themselves when they come into contact with their oppositely charged normal matter counterparts. Collisions such as these release pure energy during the collision (the highest energy photons that can be created which are called gamma rays). Gamma rays are so energetic that they can pass right through almost all matter, but they will occasionally strike an atom, transferring its enormous energy to that atom. If the atom happens to be part of a DNA molecule in some animal's body, the gamma ray can cause a random mutation in the animal's DNA (for better or worse, but usually for worse). Every now and then, however, one of these mutations gives a species an advantage, which helps it to survive in its current or a new environment.

Antiparticles are more difficult to create than normal matter because more energy is required to create an "anti" string vibration than to create their normal counterparts. This is important, because if antimatter was as common as normal matter, humans would not exist. If antimatter was as common, it would have immediately collided with the entire universe's store of normal matter right after the creation of the universe. The only thing left over from this collision would be pure photon energy, with no matter whatsoever left in the universe to create stars, planets, or human beings.

In 1980 the theories that evolved from Kaluza's 1919 theory of light being a "vibration of the fourth spatial dimension" were expanded into a competing theory to string theory called "supergravity," created by physicist Peter van Nieuwenhuizen (among others).[10] By combining our three regular spatial dimensions for gravity with a fourth spatial dimension for light the fifth, sixth, and seventh dimensions for the nuclear forces, and the eighth, ninth, and tenth spatial dimensions for matter (plus a time dimension), van Nieuwenhuizen nearly succeeded in unifying all the forces in nature into a single theory (see fig. 1.3). Unfortunately, this theory predicted types of matter that do not exist and failed to predict other known particles. Eventually, this area of research was abandoned by most physicists. Instead, many pursued ten-dimensional string theory during the 1990s. However, the theory they left behind that attempted to unify all forces into an eleven-dimensional matrix would have a kind of revenge in the end.

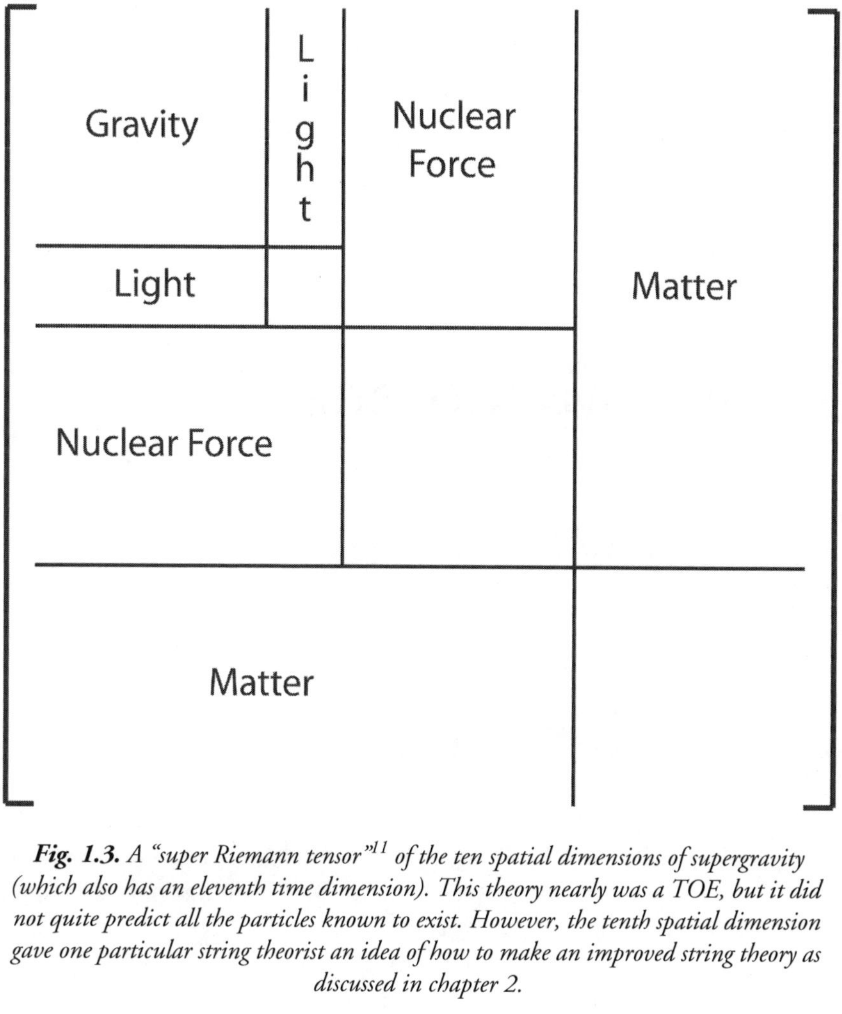

Gravity

Light

Nuclear Force

Light

Matter

Nuclear Force

Matter

Fig. 1.3. *A "super Riemann tensor"*[11] *of the ten spatial dimensions of supergravity (which also has an eleventh time dimension). This theory nearly was a TOE, but it did not quite predict all the particles known to exist. However, the tenth spatial dimension gave one particular string theorist an idea of how to make an improved string theory as discussed in chapter 2.*

CHAPTER 2

M-Theory, where "M" Is for "Membrane" (also Mega, Magic, Matrix, or Mother-of-All...)

Okay, you decide. By the early 1990s, string theory had made great leaps forward in describing the entire universe within a single framework, simply by using the shapes and characteristics of strings of energy vibrating in ten dimensions. However, string theorists were embarrassed by the fact that there was not just one but *five* different string theories, all of which seemed to be perfectly plausible. It was not until 1995 that, through much hard work and perseverance, Edward Witten developed M-theory, and string theory became a real contender for a true "Theory of Everything."[1]

M-theory suggested that the five different versions of ten-dimensional string theory were just different ways of looking at the same thing. The five different theories were similar to standing in front of a five-segmented mirror while trying on clothes. The view in each segment of the mirror is different, but the subject in each segment of the mirror is the same. Witten realized that by adding an eleventh dimension to string theory—a dimension even smaller than the other six additional spatial dimensions and the same number that supergravity had predicted a decade before—almost everything has fallen into place. Instead of a one-dimensional string vibrating in ten total dimensions, it was now a two-dimensional string (shaped similar to the outer surface of a bicycle tire's inner tube) vibrating in those dimensions, making a total of eleven dimensions.

This eleventh dimension meant that membranes, or "branes" for short, could form as these strings interacted with each other. Occasionally, a two-dimensional doughnut-shaped string could expand, flattening itself out into a two-dimensional sheet when the "doughnut hole" collapsed in the center of the string, filling up the hole, and the outer edge of the doughnut expanded into a huge, flat sheet similar to what happens when someone tosses spinning pizza dough into the air. Also, other types of multidimensional branes (the humorously named *p*-branes, where "p" is a whole number between two and nine, and one of the dimensions is time, and the others are spatial dimensions) can form as well.

Instead of all the strings in our universe being independent loops, as they were in the original string theories, M-theory specifies that most strings are open-ended like a violin string, with both ends of each string being connected to the membrane. On a flat brane, the string would bend so that both ends would connect to the brane in an arc. According to M-theory, our universe, being made up of these strings, is stuck to a huge four-dimensional membrane with three spatial and one time dimensions. Because the membrane would have evolved first, we now see strings as disturbances generated within the brane and the strings are therefore connected to the brane (more on this in chapter 8). The strings glide across or through the brane, unable to leave it or perceive anything outside of it. This is why we can only see the three spatial and one time dimensions around us and only one universe.

This theory, however, allows for the possibility that gravity is transmitted via a closed-loop string that *can* escape our membrane. It can do this because a closed-loop string has no ends to connect to a membrane, which would account for the weakness of the force of gravity compared to all other forces, such as electromagnetism. Because gravity cannot attach to or interact with our brane, it leaks from the brane, diluting its force as it can only affect other strings while it is connected to our brane. As soon as a gravity string leaves our brane, it has no way to interact with our strings, and its force disappears, giving gravity the weak characteristics we observe. As of this writing, physicists are currently searching for "disappearing gravitons" in order to verify this theory.

This new possibility of membranes also opens the door to a different way to create a universe. The membrane to which our universe is attached may be only one of a multitude of other branes of various shapes and sizes, including a four-dimensional brane-filled sphere floating randomly through our higher dimensions. As time passes, these membranes would start to bend, sway, flutter, and float around due to the chaotic quantum interactions occurring on and within them, and they could begin to pick up momentum. Eventually two of these branes would collide, releasing enough energy to create an entire new universe like our own starting at the point of the collision. This new universe would expand throughout the membrane in a similar fashion as our big bang model predicted the universe was created.

This expansion of string theory with branes, referred to as M-theory, could explain not only how our universe began but also possibly what came before its creation as well. I will elaborate more on what may have come before the creation of our universe and how everything may have started in chapter 8.

CHAPTER 3

The Motions of Isolated Volumes of Spacetime or "SlipString Drive"

Now that we have covered the basics of string theory, we can start to explore how to beat the system (or at least bend it a bit). Most theories of "faster-than-light" travel (that are based on sound scientific principles) involve wormholes, folding space, or other incredibly energy-intensive solutions. These theories often purport using a large percentage of the energy in the entire universe to create the wormhole or spatial fold. In order to create a wormhole, we would most likely need two enormous black holes or a compact neutron star and attempt to tear spacetime between the two black holes or manipulate the neutron star to spin and twist at close to the speed of light, tearing spacetime. Unfortunately, it is unlikely that the tear in spacetime will actually go where you want it to, and you will probably get stuck in the middle of the phenomenon if you attempt to fly through the resulting tear.[1] Furthermore, you will need to travel to the black hole or neutron star via conventional slower-than-light methods, making your trip impossible within one's lifetime. There have been slightly more hopeful attempts by Miguel Alcubierre and others.

Alcubierre's model proposes contracting spacetime in front of a ship and expanding it behind the ship using negative quantum energy. However, because his proposal uses negative quantum energy densities, it is highly unlikely to work in the real world. Such energy may not even exist on the macroscopic scale or be usable on any large scale applications as it is a quantum effect. Other problems include not being able to control your movement from within the bubble or to stop your movement once started.[2] Unfortunately, because of these drawbacks and more, theories such as Alcubierre's are not very likely to happen on any practical sort of level, even in the distant future.

Most other recently proposed methods of faster-than-light travel involve backward time travel, which is more than likely impossible (sorry). Once the laws of physics are unified (for example with a *Membrane Theory of Gravity* described in chapter 8), time should simply be shown to be the rate at which strings vibrate. This rate can be slowed down due to relativistic effects but not made to vibrate in reverse to any significant degree, invalidating all theories that involve backward time travel. There are, of course, many science-fiction methods of faster-than-light travel. But as with *Star Trek Voyager's* similarly named "Slipstream drive," they involve travel through either subspace or hyperspace, which do not exist in the real world.

Other fictional methods of rapid space travel simply do not obey the laws of physics. This includes the 2003 novel by John Varley, *Red Thunder*. The book includes a "squeezer" power source which provides limitless power for the ship, which violates physics and thermodynamics (nothing can produce limitless power). Apparently, the device uses something similar to "zero point energy" to produce its power. As we will read in chapter 8, zero point energy probably does not exist either. The current acceleration of the expansion of the universe is not because every point in space has a repulsive energy, but is due to the *membrane theory of gravity*. Such acceleration will not last forever. Ideas such as vacuum energy are being promoted by some physicists and on TV shows such as *Stargate SG-1* because physicists don't yet know what dark energy is, leading to wild speculation on the subject. No one has ever detected vacuum energy, let alone actually figure out a way to use it. Both dark energy and dark matter will be explained in chapter 8, putting to rest these ideas about the availability of massive amounts of vacuum energy to power ships (or anything else for that matter). Due to the problems with other proposed methods of "faster-than-light" travel, I believe only SlipString Drive has a realistic chance of success in the near future. It is the only scientific method based on the use of gravity waves. Gravity waves are a fundamental part of Einstein's Theory of Relativity (and every part of relativity that has been tested has been confirmed). SlipString Drive has significant advantages over the possibility of wormholes (or all the other methods mentioned above) for interstellar travel, and the technology might even be practical within the next fifty years or so.

First, let us explore why we cannot conventionally travel faster than light. As with relativity, the closer you get to the speed of light (C), the more massive you become, and the more energy it takes for you to accelerate closer to C. Additionally, the closer you get to C, the more that time will slow down for you relative to the rest of the universe. If you or any other matter did happen to actually reach the speed of light, your mass would become infinite (really, it is time to go on a diet!) And time would slow to a complete stop for you and possibly even for the rest of the universe. Obviously, this scenario is quite impossible.

The consequence is that by using conventional methods (rockets, solar sails, ion drive, plasma drive, etc.) and giving yourself the highest benefit of the doubt (as far as your speed is concerned), it would take you a minimum of nearly five years to reach Alpha Centauri, the closest star system to us. (Current technology could take nearly forty to fifty years, but let us say you were able to travel quite close to the speed of light.) Alpha Centauri is 4.3 light-years away, and if you were traveling extremely close to the speed of light for most of your journey (which is highly unlikely but good for an example), only about one year would pass for you while nearly five years pass for the folks back home on Earth (depending on your top speed and how long you stayed there). Another side-effect of traveling that fast involves dust particles becoming lethal projectiles as they strike your ship at close to the speed of light, turning the ship—not to mention your passengers—into something resembling Swiss-cheese. This is not a very practical way to get around our universe in a hurry (and none too healthy for your passengers, either, especially if you like to travel in one piece!).

Explaining the reasons behind these side effects is difficult. Physics can describe these effects and what will happen but not exactly *why* these effects occur. However, here is one possible answer to why they occur. Let us imagine that you are pushing your finger on a piece of fabric (similar to spacetime). The harder and faster you push your finger forward, the more the fabric bunches and folds up in front of your finger and the more difficult it becomes to push your finger further forward. This is similar to approaching the speed of light in a traditional spaceship. The closer you get to the speed of light, the more the strings that make up the universe start getting compressed in front of your ship. Actually, it is the membrane that contains our universe that most likely gets compressed (due to the *Membrane Theory of Gravity*), but let us continue. This creates more and more resistance, making it increasingly difficult to accelerate any further (and also increasing your mass as you start to drag more of the rest of the universe along with you). Time will start to slow down from your perspective as well. This can be understood by imagining the strings that make up spacetime squeezing back in on your ship from all directions with greater and greater force the faster you attempt to travel. The inward pressure of spacetime becomes so great that it compresses the strings that make up your ship and yourself. This prevents the strings in the atoms of your ship and body from vibrating at their normal speed because they are being compressed so tightly and have virtually nowhere to move. Therefore, the closer you get to the speed of light, the more that time slows down from your perspective as the strings within your atoms vibrate more and more slowly.

Another perspective is to think of space and spacetime as a fluid, albeit a very thin fluid. Einstein predicted a gravitational effect called "frame dragging," and NASA recently launched the Gravity Probe B satellite with the highest precision gyroscopes ever manufactured on board to attempt to prove Einstein's theories of gravity. It has since reported home and scientists are currently verifying the data in order to prove both frame dragging and the geodetic effect.[3] Frame dragging is the effect that massive bodies have when they spin in space. As these objects spin, they drag spacetime along with them, so that the space around a planet or star actually twists along with the rotation of the body (very, very slightly with planets like Earth without much mass). This effect is similar to a ball rotating in a bucket of water. As the ball spins, the water around it picks up some of the momentum from the ball, and the water (space) starts to spin along with the ball. Gravity Probe B is also attempting to prove that spacetime is warped by massive bodies (the geodetic effect), which accounts for the "missing inch" as a satellite orbits Earth (relative to Newtonian physics). So the distance a satellite travels is one inch shorter than would be calculated using Newton's laws, because gravity warps spacetime around massive bodies.

During final editing of this book, the equations of general relativity were confirmed to be accurate within 0.05 percent. [4] They were validated in a study of binary pulsars between April 2003 and January 2006 by an international team led by astronomer Michael Kramer of the Jodrell Bank Observatory in Macclesfield, U.K. Binary pulsars are akin to spinning atomic clocks in space and their rapid rotation, orbits, and massive gravity cause large relativistic effects such as curving and twisting spacetime. Due to these factors, the team was able to decipher the twin pulsars' radio waves and calculate the stars' orbits to a very precise degree, confirming general relativity to extreme precision including the effects described above. This finding puts pressure on quantum mechanics as it must be able to incorporate general relativity to describe the universe, and can not currently do so.

Star Trek's "Warp Drive" was almost on the right track for faster-than-light travel, but they probably got it backward. If they warped spacetime inward toward their ship, they would actually *increase* their mass and, therefore, cause all the relativistic effects on their ship to get even worse, making it much harder to travel faster! However troublesome the physics of *Star Trek* may be, the show inspired me and a large segment of the population with its optimistic message about the future of humanity. I hope this book will also inspire its readers about the possibilities the future holds.

Before developing SlipString Drive (soon after first reading about string theory) I originally considered that one might be able to "cut" one's way through the universe. By generating a particular frequency or amount of energy along the hull of a ship, perhaps there was a way to slice through strings, repel them, or make them unstable in some way. By doing this, your top speed would no longer be C but the rate at which you could "slice" through the universe. Unfortunately, as I continued reading, it soon became obvious that the likely energies required to make even one string unstable are far beyond our current (and probably future) capabilities, so there went that idea. However, it did make me think that if there was some way to isolate ourselves from the rest of the universe, it might be possible to travel almost as fast as one wanted. After much thought, I finally realized that gravity waves might hold the answer.

Gravity waves actually have a repulsive effect on matter and on the rest of the universe in general. This is supported by relativity, which states if our sun were to instantly disappear, the planets would continue to orbit the sun's original position until the gravity wave produced by the sun's disappearance traveled outward (at the speed of light) and crossed each planet's path. When the gravity wave reached each planet in turn, the planet would be pushed outward just enough so that the wave cancelled out the gravity that the sun had generated before it disappeared. The planets would then travel in a straight line from the moment the gravity wave crossed their path, and the planets would then be released from the sun's gravity well (see fig. 3.1).

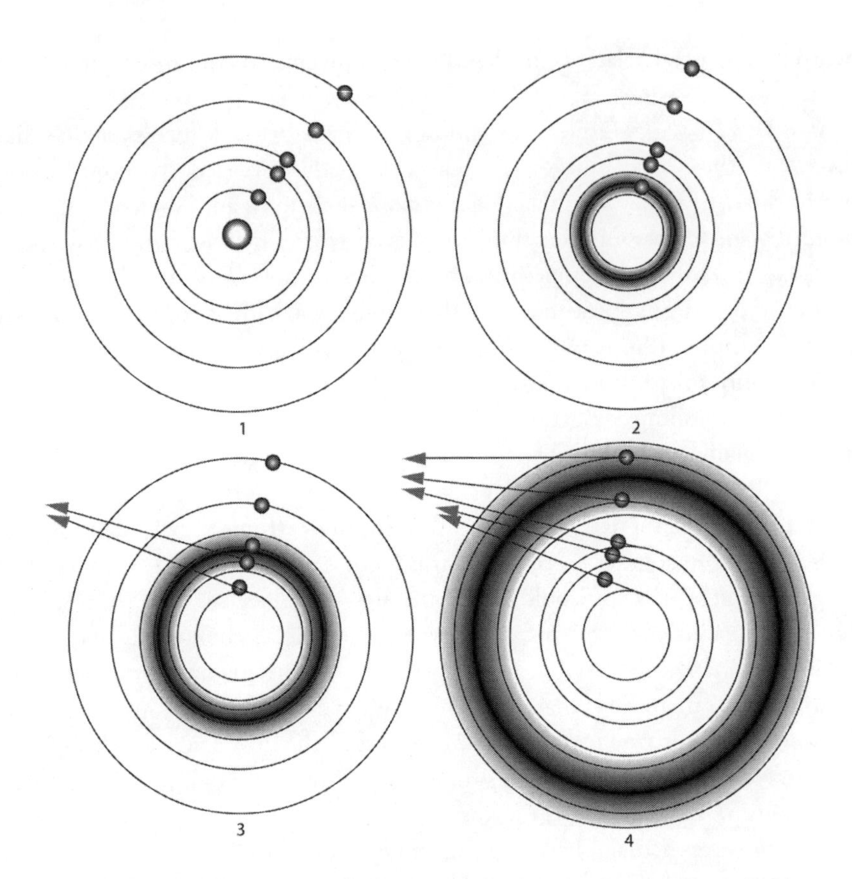

Fig. 3.2. *The effects and motion of gravity waves. (Speed and scale are highly exaggerated.) If a star (1) were to suddenly disappear, (2) the planets orbiting it would continue following their orbits until the gravity wave of compressed spacetime (released from the star's location) passes each planet (3 and 4). Spacetime had been compressed by the mass of the star, which bends light waves and the space around it due to its huge mass. When the star disappears, the spacetime that had been compressed by the star is released from its hold, and a wave of compressed spacetime rushes outward from the source at the speed of light. This wave pushes the planets away precisely as much as the star's gravity used to keep the planets in their orbits. As the wave crosses each planet's path in turn, the planet then travels in a straight line, released from the star's gravity. As the gravity wave's speed is limited to the speed of light, it crosses the path of the next planet farther out between several minutes and many hours later than the previous one. The farther out a planet is, the smaller the angle at which it will leave the solar system.*

A gravity wave is a compression of spacetime, and that compression wave radiates outward from its source (such as a supernova: a massive star that explodes at the end of its life). This has the effect of pushing spacetime (and anything in its path) away from the wave's source, similar to how a water wave pushes against a rubber duck. Unlike a water wave that gets taller with the more energy it is given, a gravity wave compresses a larger amount of spacetime into a smaller area with the more energy it is given. The wave always travels away from its source at the speed of light (the speed at which gravity and light waves propagate throughout our universe).

The Motions of Isolated Volumes of Spacetime

At this point, I would like to discuss the motions of "isolated volumes of spacetime." These volumes can be generated by a source emitting gravity waves in all directions and by various amounts. To external viewers, the motions of objects within these isolated volumes may appear to violate certain laws of physics. However, as you will see, they do not violate any laws of physics if viewed from the proper relativistic perspective.

There are two types of isolated spacetime:

1. Gravitationally isolated spacetime. By means of using gravity waves emitted in all directions from a gravity wave source, enough gravity waves are produced to negate the mass of the source, effectively reducing the mass of the isolated region to zero. However, because of gravity's weakness, light waves and matter can still pass through this partially isolated region to the rest of spacetime and vice versa. Light waves may be slightly distorted due to the gravity waves generated to produce this region, and matter would curve away from this region, being repelled somewhat by the gravity waves but still able to pass through if given enough momentum.

2. Completely isolated spacetime. By means of using gravity waves emitted in all directions from a gravity wave source, enough gravity waves are produced and caused to constructively interfere creating a standing wave in order to completely "pinch off" and isolate a volume of spacetime from the rest of our universe. No energy or matter whatsoever can pass between the isolated region and the rest of our universe.

There is no physical reason these isolated regions cannot be created if sufficient gravity waves can be generated in order to form them. It may be easier to visualize these waves in a membrane gravity universe as proposed in chapter 8, but traditional relativistic gravity illustrated by the two-dimensional diagrams in fig. 3.2 will work just as well. Additionally, some readers might think that by effectively reducing the source's mass to zero through these processes that this might violate an equation such as $E = MC^2$. No physics pertaining to gravity is violated by these processes because the source is within a gravitationally isolated region of spacetime. Just because the source's mass appears to be zero from an external observer's point of view does not mean that its mass actually is zero. The source's mass stays constant, but the region of spacetime it occupies is now gravitationally isolated from an external observer. This causes the observer to believe something is amiss when actually no violation of physics has occurred. Similar to relativity, it all depends on your point-of-view and whether you are inside or outside one of these isolated regions.

Fig. 3.2. *A two-dimensional representation of gravity waves being emitted from a source. Using constructive interference as gravity waves overlap each other and by generating massive amounts of gravity waves, the region the source is within could be "pinched off" from the rest of spacetime, isolating it from the rest of our universe. The opposite of an event horizon is generated: a point at which spacetime is so curved that no light or matter can enter or leave the isolated region of spacetime, cutting it off from the rest of the universe. This illustration is not completely accurate as no waves would pass beyond the largest peak, and external spacetime would be pushing back in on the region as energetically as the source is pushing out. Instead of peaks and troughs, think of gravity waves like this: Where you see a peak, a larger amount of spacetime is more densely compressed into a smaller volume, and where you see a valley, a small amount of spacetime is stretched very thin. This is similar in fashion to how sound waves propagate through our atmosphere.*

By generating a gravitationally isolated region of spacetime, the source of the waves is now able to move and accelerate at speeds that could appear to violate the laws of inertia by an observer outside the region. However, from within the gravitationally isolated region, no motion whatsoever is occurring. The region itself can be moved by means of generating additional gravity waves and aiming them in the opposite direction of the intended motion. If maneuvering within a gravitationally isolated region only, local gravity sources must be taken into account. If there is a large gravity source such as a planet nearby, additional gravity waves must be produced in order to cancel out the gravity of the external source and prevent falling into the object's gravity well. If a gravitationally isolated region is created or lands on the surface of such a massive object, excess gravity waves must be generated toward the object in order to overcome the object's gravity and to escape from the object's gravity well.

By generating a completely isolated region of spacetime as in figs. 3.2 and 3.3, the source of the waves is able to travel using the curvature of spacetime itself to move from one point in spacetime to another. Once a region is completely isolated, the outward pressure of the gravity waves causes the spacetime outside the region to press back in on the isolated region with equal force. This pressure can be used to propel the region from one point in spacetime to another. By generating additional gravity waves on one side of the source, this will deform the spherical bubble of the isolated region into a teardrop shape, causing external spacetime's inward pressure to be distributed unevenly around the isolated region, forcing it rapidly in the opposite direction (see fig. 3.4). To the dismay of many a physicist (and making myself quite uneasy when I first pondered the idea), this motion is no longer limited by the speed of light. The isolated region's speed is not the result of motion through spacetime but rather of one isolated bubble of spacetime being propelled by the rest of spacetime, squeezing the isolated region forward. The isolated region accelerates without resistance, and there is no method of generating relativistic effects to slow this acceleration. The external region of spacetime squeezes the isolated region forward similar to how a magnetically levitated train moves but with no air resistance or relativistic resistance whatsoever, allowing it to slip from one point to another, effectively bypassing the rest of the universe. This type of motion is similar to traveling through a wormhole in that it is simply taking a shortcut around spacetime rather than flying directly through it using a brute-force method. For a more down-to-earth example, it is similar to squeezing a slippery watermelon seed between your thumb and index finger, causing it to rapidly shoot out from between your fingers due to its teardrop shape and slippery nature.

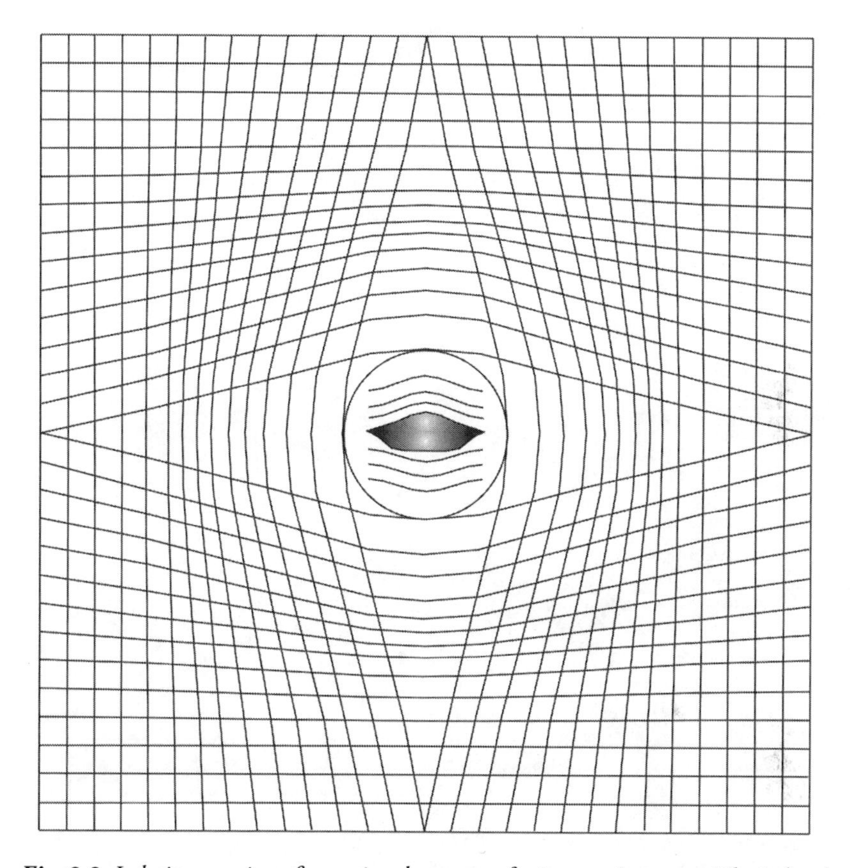

Fig. 3.3. *Isolating a region of spacetime by means of using gravity waves. The isolated region of spacetime is no longer in contact with the rest of our universe, allowing it to move independently of external spacetime. In fact, an external observer would see the source "disappear" as its gravity wave bubble increases in strength and eventually completely envelops the source. Light waves will be bent farther and farther away from the source until it is completely within its own isolated region of spacetime and is no longer visible from outside the region.*

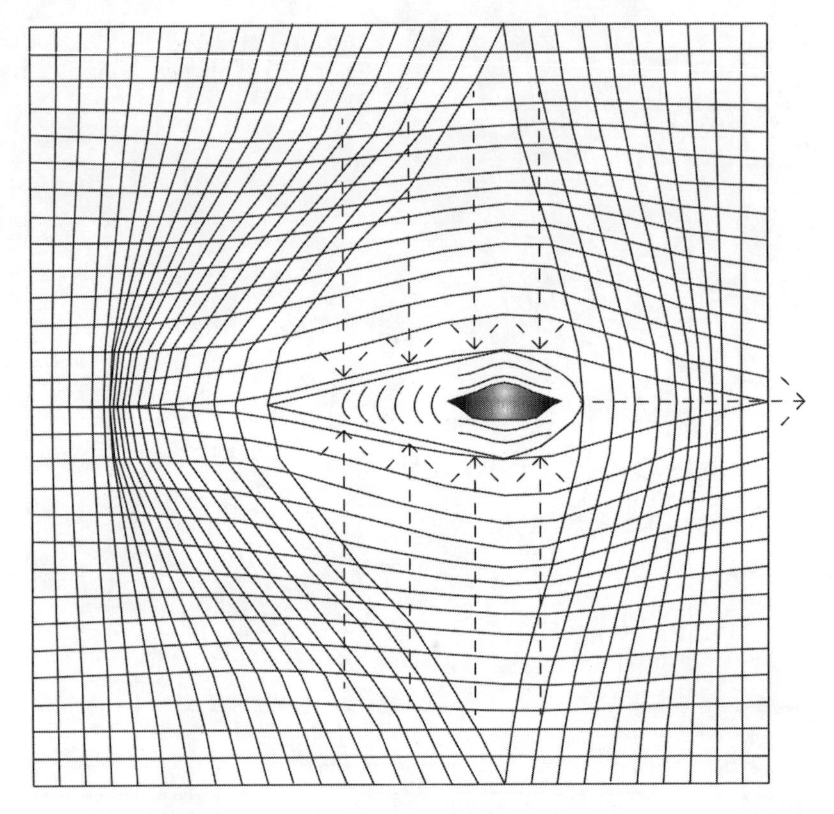

Fig. 3.4. *How to accelerate an isolated volume of spacetime by changing its shape. By generating additional gravity waves behind the source, this action deforms the shape of the isolated volume of spacetime. By deforming the shape, this causes the external region of spacetime to compress the isolated volume unevenly (more behind the source than in front of it). This uneven pressure causes the isolated volume to be accelerated in the opposite direction of the deformation at extremely high velocities and without resistance, as the rest of spacetime glides around the isolated region's gravity wave bubble.*

As opposed to traditional motion through spacetime, which increases mass and decreases the passage of time the higher the velocity of the traveler, travel within a completely isolated region of spacetime has no intrinsic limitation on speed and no time penalties for increased velocity. The passage of time within the isolated region stays constant no matter what shape you make it or how fast it travels. There is no resistance to the motion of a completely isolated region that would increase the mass of the region or cause any relativistic effects to slow or alter time within the region. The only factors governing its motion are the amount of acceleration due to the degree of deformation of the region and the length of time the region spends accelerating and decelerating. (Deceleration is produced by the deformation of the isolated region by an equal amount in exactly the opposite direction of its acceleration.) Because the isolated region is effectively massless, the immense pressure of spacetime propelling this region will cause it to accelerate at unheard of velocities. Therefore, depending on this rate of acceleration, speeds of many thousands, or even millions, of multiples of C could be achieved in quite a short period of time with no ill-effects or relativistic penalties to its occupants. Once deceleration is completed, the source simply reduces the strength of the gravity waves generated in all directions, and the region will come back into contact with the rest of the universe near its destination.

Of course, travel using this method of propulsion will create several challenges. From within a completely isolated region of spacetime, the source would be "as blind as a bat," for there is no way to see anything outside of the isolated region. Even with this difficulty, however, one could easily figure out a way to calculate the acceleration of the region either through computer models or by running test trials. These trials could be executed by increasing the gravity waves behind the source for a short, specific amount of time and then by equalizing the bubble. Next, the source will slowly reduce the gravity waves generated around it until the region is slightly back in touch with the rest of spacetime, and then measure the distance the source has traveled and its current velocity. With these measurements, it would be possible to calculate the acceleration rate and how long it would be necessary to accelerate and decelerate to arrive at a particular destination. Once a pilot has chosen a direction in which to travel and knows the rate of acceleration, the pilot can completely force spacetime outward, stretch it, and accelerate as rapidly as possible toward the target solar system or other point of interest.

Imagine your finger pushing on a piece fabric again. Instead of the fabric (strings) bunching up in front of your finger whichever way you go, you now have a string-repulsing blade. Set it on the fabric, and slide it in any direction. As you slide the blade across the fabric, it has no resistance whatsoever. The fabric simply opens up and spreads apart around the blade. In this manner, we can now "bypass" spacetime and therefore the speed of light, as well as all relativistic effects. As far as the rest of the universe is concerned, the source does not even exist when completely isolated within its own volume of spacetime.

While rapidly propelling a craft in this manner, there is no longer a need to worry about one's twin brother or sister being fifty or one hundred years older when returning from a long journey. This is so because the strings (spacetime) within the isolated region are still vibrating at roughly the same rate as the strings that make up most the rest of the universe and the folks back home. The standard rate of vibration of the strings that make up our universe is constant. However, relativistic effects can slow them down more in different areas of the universe due to massive gravitational fields or very rapid movement through spacetime. If a traveler is completely isolated within a region of spacetime, the inside of the ship is not being compressed by gravity any more than it would be outside the region. Because the gravity waves are being emitted from the hull of the ship, they have no effect on the ship's passengers. Also, the problem of dust particles turning one's ship and passengers into Swiss-cheese at speeds close to C is no longer an issue. These dust particles simply stay put exactly where they are as the space they occupy slides around the gravity wave bubble, not even realizing the region passed by. In this manner, we break no laws of physics. The speed of light is still unthreatened as the speed limit for the rest of the universe. Our region of completely isolated spacetime is simply no longer connected to the rest of the universe!

It would be necessary to create a precise computer model of a ship propelled in this manner in order to navigate while within an isolated region. The model should include the gravity waves the ship produces, the gravitationally and completely isolated region's physics of acceleration, the ship's design, and an accurate star map. The map would need to be updated regularly with as much information as possible on surrounding star systems, nebulae, black holes, and other astronomical objects. With a computer model such as this, you could see (on your computer screen) your position, speed, acceleration, etc. as you travel toward your destination (even if you cannot actually see outside your isolated region during the trip). You would need to be extremely careful and leave plenty of distance between your ship and the target solar system when exiting the isolated region. You would not want to accidentally come out of SlipString Drive in the middle of an asteroid belt or gas giant. It would be prudent to have the computer automatically note any newly discovered objects in the ship's computer model as you explore. This would allow you to travel more accurately in the future and to avoid any major hazards or collisions. A central database on Earth could be used to receive updates from all ships so that after returning home each ship could automatically update its database with the latest charts in order to stay safe while traveling.

A ship designed to travel in this manner would need to be completely covered with electronic photon sensors so that you can see clearly after coming out of SlipString Drive. While darting around in one of these ships exploring a planet, you only need to generate enough gravity waves to cancel out your ship's mass and to maneuver (creating a gravitationally isolated region only). This smaller use of gravity waves will not distort light waves very much. However, you would want the ship's computer to adjust the view screen or images projected inside the ship (like a heads-up display) to correct for the gravity wave distortions produced by the ship's engines. The computer would simply read the strength of the gravity waves the engine produces and bend the images enough to cancel out those distortions. This program would also be useful as you prepare to go into a completely isolated region of spacetime. The more you increase your gravity wave bubble, the more the program would need to take the shrinking area visible outside of the ship and spread it out visually in front of you. This way you can see clearly until everything disappears when your gravity waves completely isolate the ship.

If you wanted to be able to see outside of your isolated region of spacetime while still safe within it, we could create a periscope to accomplish this. The periscope would stick up from the top of a ship and would be the first thing to emerge from the isolated region as the gravity wave bubble was decreased. This way the only part of the ship outside of the bubble is the tip of the periscope, and it is the only part vulnerable to external forces (see fig. 3.5). This would create a way to make sure that you are safe before exiting the security of your isolated region of spacetime. This would have the added benefit of allowing covert observation of others without giving away what and, possibly, where you are. At most, a ship might appear to be a small glowing ball due to exciting gases in the atmosphere. The periscope and parts of your ship may glow possibly because of the high energies created by the engine powering your ship (unless an efficient method of producing gravity waves can be discovered). In space, it might appear to be a small sphere about the size of a soccer ball or less as no gas would be excited to call attention to the ship.

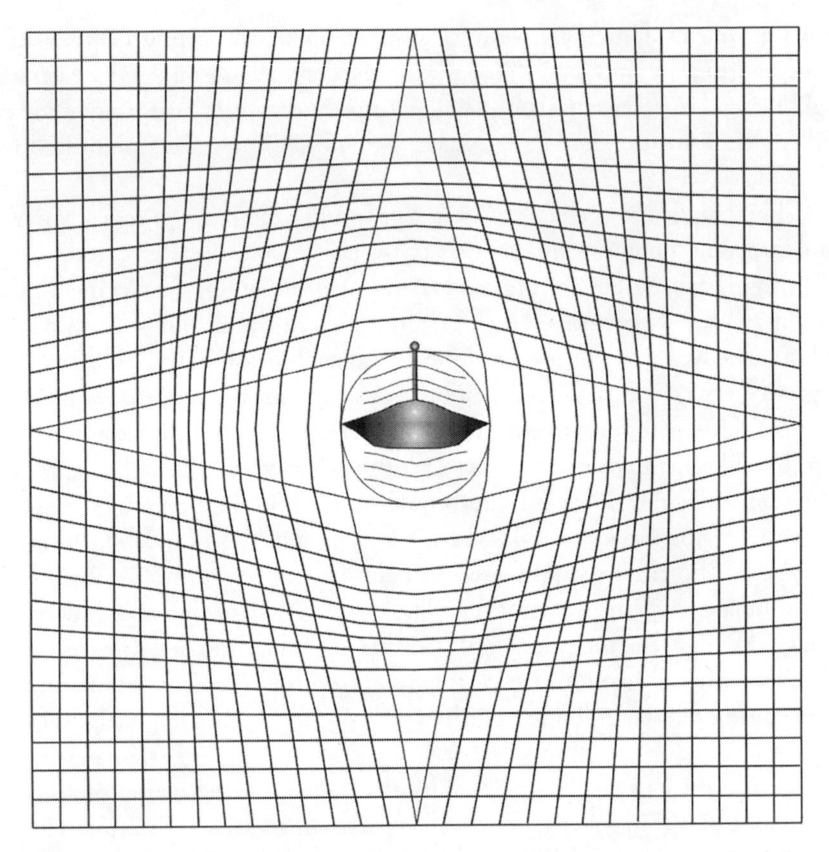

Fig. 3.5. *Using a periscope to see outside of an isolated region of spacetime. As a pilot reduces the gravity waves isolating the ship from the rest of the universe, the pilot could see outside of the isolated region without leaving the ship vulnerable to external forces. If a pilot exits the isolated region near a gas giant or black hole, etc., the pilot risks destroying the ship if the ship exits too close to it. Although this may appear somewhat silly, a periscope such as this would have the added benefit of giving a pilot the ability to observe others outside of an isolated region without being vulnerable to external forces or revealing what the ship is. External observers would see only a small sphere, which could be small enough to be ignored by radar or similar technologies, especially if hovering over a technologically advanced planet.*

If you were piloting one of these ships and had the misfortune of navigating it into the middle of a gas giant after a long trip, at least with a periscope up, you would know that you had better not burst your bubble in that location! A periscope could be used like training wheels until a pilot became proficient at traveling using SlipString Drive and could also be used as a safety feature even for experienced pilots. It could also be useful for observing other species and planets without being observed, especially from space. Your periscope would need to emit some gravity waves in order to balance out your field as you traveled, but you could reduce those waves around the tip of the periscope in order to observe outside your isolated region.

In general, try not to think of the motions of completely isolated regions of spacetime as traveling faster than light but simply as warping spacetime to an extreme degree (and taking advantage of curved spacetime as a wormhole does). This theory is similar to relativity in that it is all about one's point of view. If it were possible for someone to watch a ship propelled in this manner from outside of a completely isolated region, the ship would appear to be doing all sorts of impossible things. However, from within the completely isolated region, a pilot could not even tell that the ship was moving at all without the help of a computer's projections.

Try not to take physics too literally, as Stephen Hawking did for a short while in his best-selling book, *A Brief History of Time*. Hawking originally thought that time would flow backward if the universe started collapsing into a big crunch. A big crunch is what would happen if there were enough mass in the universe for it to overcome its expansion. This would cause the universe to stop expanding and to collapse back in on itself, shrinking back down to a singularity, and possibly starting the universe all over again. As with a big crunch, time will not reverse when traveling faster than light using this method, partly because backward time travel is most likely impossible, and also because it would be reading physics too literally.

One cannot apply equations such as $E = MC^2$ to this type of travel. First of all, C is a constant, and the equation would be meaningless if C were a variable. Secondly, that particular equation was not intended to be used under these conditions but only to calculate the amount of energy in a given mass. Because we do not yet have a comprehensive theory of everything, our current equations apply only to specific circumstances, such as measuring the amount of energy in a given mass. However, relativity should be sufficient to describe these isolated regions of spacetime on its own. For those who do know relativity, when traveling in this manner you are just "expanding the cone of your world line to higher than 45 degrees in your Minkowski diagram, putting more spacetime events in your cone of possible pasts and futures and shrinking 'elsewhere' down."[5] It is unsettling to us as physicists because we have been drilled from day one that C is a constant and that travel faster than light is impossible (which is still true for any brute-force methods of attempting such travel). We just need to recognize that there are ways other than brute force to effectively bypass long stretches of spacetime by means of using completely isolated regions of spacetime.

CHAPTER 4

UFOs as Proof-of-Concept? (No, I am not trying to prove they exist!)

This chapter might get me into a little trouble. I want to be as credible as possible in writing this book, and I certainly do not want to be labeled a UFO nut. For a long time I have believed that UFO theorists had little credibility and that the universe was a nice place with universal laws that could not be bent or broken. I believed that things such as UFOs were highly improbable (to say the least). If this hypothesis for SlipString Drive is proven to work, however, it would make it possible for intelligent species from other planets to visit our solar system, but I am not going to comment on how likely that scenario might be. I still have many issues with most of the TV shows and "UFOlogists" out there trying to claim that UFOs do exist, plus the fact that there have been quite a few obvious fake photos and videos from people trying much too hard to get some attention.

One complaint of mine is that some TV programs still try to use the "Face on Mars" photo, taken by Viking Orbiter I in 1976, as proof that intelligent life exists outside our planet, completely ignoring the fact that the last Mars orbiter revisited the site. The newest photos are at a much higher resolution and without as many shadows, revealing the face to be the small rocky range it really is and nothing like a monument. The appearance of a face was simply the result of the low-resolution images on the earlier Mars probe as well as some active imaginations on this planet filling in the gaps in the data. I could claim that a Mars crater looks like Kermit the Frog, but that does not mean I believe that "Pigs In Space" was anything more than a good skit on *The Muppet Show* or that aliens would pay homage to it by carving Kermit's likeness into the terrain. I am sure they would have much more productive things to do with their time if they exist.

Another gripe I have is how alien encounters are almost always portrayed in movies and on TV: The aliens are hostile, trying to take over the earth or the rest of the galaxy (if not the entire universe). If alien species do exist and have the technology for interstellar travel as proposed in this book, it is highly unlikely that they will need to expand to other planetary real estate and especially not to any other inhabited world. Advanced civilizations probably have virtually limitless fusion or antimatter power, manufacturing capabilities, food (and/or a genetically engineered need for much less of it), and no need to reproduce any more than just to keep the population constant. They would not want to overpopulate themselves, because they know that huge numbers of their species (in the billions) are just not sustainable in the long term. Most likely, after millions of years, they will have moved toward becoming an almost exclusively space-faring race. Why would they live on a planet that is a huge target for any number of natural disasters and will eventually die when its sun does, if not long before? Instead, they can build huge self-sustaining city-ships in space that would be immune from asteroids, comets, or a sun at the end of its life. They could even be protected from a radiation wave from a supernova by using heavy elements to shield the city-sized ship or by creating an isolated region of spacetime around the ship and simply letting the wave pass around the region.

If an alien species for some reason did need more room to expand, however, it would be so much easier for them to simply settle on an uninhabited planet somewhere else. They could settle on a planet similar to Mars or on any of hundreds of millions of other planets without intelligent life on them. Why waste incredible resources fighting intelligent beings off on another planet when an alien species can simply move to another uninhabited planet with virtually no effort whatsoever! However, this scenario does not make for a very exciting movie or TV show, so here we go again with those nasty, hostile, power-hungry aliens. A species intelligent enough to travel the cosmos by SlipString Drive would not need to be taking over other planets. They are smart enough to create everything they need for themselves or to simply trade with other advanced space-faring civilizations for what they do need.

On the contrary, alien species would probably think that life-sustaining planets and their possibly intelligent inhabitants are the most precious resources in the universe and would want to keep those planets and species as pristine as possible. Other beneficial plants or animals may evolve on those planets that could be of great use to an alien species' medicine or other technologies. They may want to take a sample of a different plant or animal species or a type of technology now and then but never destroy any species on an alien planet. Alien species would simply be shooting themselves in the foot if they went out and destroyed any life on another planet. Besides, one day a species from one of those planets may evolve to become a great trading partner in science, culture, and technology, having thought of ideas and concepts that the other species' brains were not quite built for. An alien species would most likely be more concerned with our own hostile nature. They would prefer we did not wipe ourselves out before being able to move on toward exploring the stars and becoming a peaceful space-faring species along with the rest of our galaxy's SlipString Drive-capable species.

Regarding UFOs, obviously I cannot prove or disprove their existence, and I am not going to try, but let us just say that many credible people including many civilian and military pilots, radar operators, and flight control staff have seen some pretty weird things. Unfortunately, I will probably be labeled a heretic anyway for suggesting that we can get around the speed of light (although I am not actually suggesting that any physical laws of the universe need to change to do so) but here goes anyway…

These are some descriptions of UFOs that people who claim to have seen them use to describe them:

1) They are silent.

2) They can pull off maneuvers that put our airplanes to shame, such as going from flying straight ahead to making a 90° turn without banking or slowing down, or they can accelerate from zero to twenty thousand mph in a half second.

3) They generally seem to "defy the laws of physics."

4) They sometimes appear to change shape and have been seen to vanish (and not just because they are flying away at high speeds).

5) They bounce, jitter, blur, and doggone it, you just cannot take a photo of one in focus!

Again, I am not saying these things actually exist or are necessarily alien, but as a thought-study, if you think about flying around in a SlipString Drive ship, it would be:

A) Silent, because gravity waves do not make any noise or disturb the atmosphere in any way that would.

B) You would be able to pull off maneuvers like turning at 90° angles without banking, or going from zero to twenty thousand mph in half a second without getting squished. You can accomplish these feats because inertia does not affect your ship. It does not because you are at the center of your own gravitationally isolated region of spacetime produced by the gravity waves emanating from your ship's hull. Therefore, you can maneuver virtually however you like (using additional gravity waves) with no regard for inertial forces, and easily maneuvering around large gravitational forces with the use of additional gravity waves. If you are not at least partly inside a gravitationally isolated region and are just running on standard gravity-wave drive (something that is definitely not recommended), then you risk getting squished by inertial forces when you try to accelerate. The relatively small amount of gravity waves needed to gravitationally isolate your ship may not even be enough to bend the light waves around your ship very much (if you do not want to), so you have a fairly clear view of your surroundings.

C) You can generally "defy our laws of physics" because until now nobody thought of the possibility of being at the center of your own repulsive gravitational field or that, if you gravitationally or completely isolate yourself in your own little region of spacetime, outside gravitational forces need not apply to you. The laws of physics are all still intact; it is just the perspective that has changed. The laws of physics just appear to be bent or broken from an outside observer's perspective. This is very confusing until you finally figure out what the trick is behind how a ship can curve spacetime enough to isolate the ship gravitationally and allow it to maneuver in seemingly impossible ways. As Arthur C. Clarke said in *Profiles of the Future,* "Any sufficiently advanced technology is indistinguishable from magic."

D) If you increase your gravity wave bubble while hovering in front of an observer, you will curve spacetime out away from your ship to such a degree that it causes light waves to bend completely around your isolated region of spacetime. Your ship will appear to shrink as your gravity wave field increases, and then your ship will completely disappear from anyone's point-of-view outside of your isolated region. This causes outside observers to believe that you have disappeared even though you are still hovering right in front of them. However, you are now in your own completely isolated region of spacetime, preventing detection by anything outside your isolated region.

E) If you are trying to photograph or videotape one of these SlipString Drive ships hovering above the ground, it will bounce around a good bit as it passes over spots where the earth's gravitational field varies (as the density and composition of metals and rocks in the earth's crust change depending on your location). Additionally, it would be quite difficult to get a photo of one of these ships completely in focus because they are using gravity waves to propel their ship! Light waves will be bent by the gravity waves that allow the ship to maneuver. If a pilot increases the strength of the gravity wave bubble around the ship to reduce the risk of hitting anything or getting hit (as outside objects are pushed away from the gravity wave field), observers will see a blurred, smaller, and distorted image. They will just see the very top, bottom, and sides of the ship getting squished closer together (see fig. 4.1), or they will see just the top and bottom of the ship squished together, possibly with a blur in-between, depending on what axis the gravity waves are being generated. You might generate gravity waves mostly along the z-axis (up and down) in the shape of a dinosaur egg in order to simplify your engine design and reduce alignment errors with your gravity waves. If you do, the top and bottom of your ship will appear to get compressed together (see fig. 4.2). There may be a blur in the center of your ship between the very top and bottom as the light waves coming from the ship are bent the most in the center where the effect of the gravity waves are strongest. The more you increase the gravity-wave bubble on one axis or on all three, the less of your ship someone can see from outside of your gravitationally isolated region.

Fig. 4.1. *What an outside observer might see as a SlipString Drive ship enters an isolated region of spacetime using a symmetrical gravity wave bubble. If a ship (shaped like a traditional rocket) is hovering over a planet, it may appear as any one of the three shapes (or something in-between) depending on how safe the pilot feels. If a pilot does not want to risk colliding with indigenous aircraft, the ship may look closer to the image on the right or smaller, well within the isolated region. The pilot would still be able to observe outside the ship using sensors on the hull. A computer program would compensate for the warping of spacetime by stretching out the pilot's view screen image in proportion to the amount that the pilot is warping spacetime.*

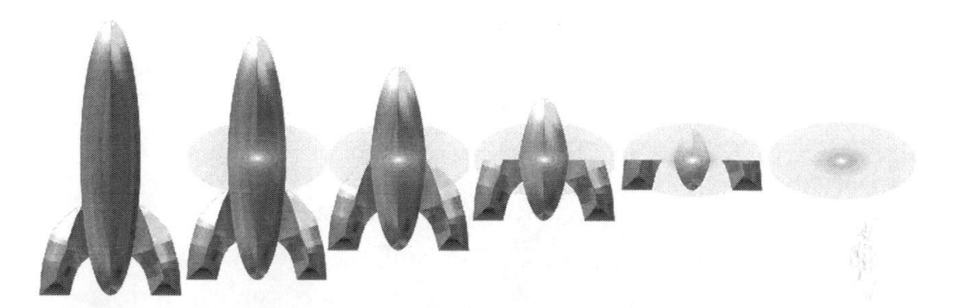

Fig. 4.2. *What an outside observer might see as a SlipString Drive ship enters an isolated region of spacetime using an asymmetrical gravity wave bubble. If a ship (shaped like a traditional rocket) is hovering over a planet, it may appear as any one of the shapes above (or something in-between) depending on how safe the pilot feels. The ship above uses a drive that diverts its gravity waves mostly along the z-axis (up and down) in a tall, thin, dinosaur egg-shaped bubble. This may be done in order to simplify engine design and would result in the ship appearing compressed from top to bottom.*

So, it could be possible that some people have seen evidence that this theory actually works. It also may be possible some day to prove that gravity waves are being used as propulsion if we see videotapes of an object that bends the light around it and appears to change shape in a way that is consistent with this hypothesis. The amount of gravity it would take to bend light in the manner I describe would be so immense that it would be almost impossible to fake a tape like this here on Earth. The only way to fake that sort of effect now is with a 3D computer model. If we find someone's old home videotape (jittery and blurry as usual) with the object bouncing in sync with the background (and not a smooth computer-generated Spielberg special effect), it is possibly a gravity wave-propelled ship and no longer a completely unidentifiable object. Atmospheric effects such as heat waves from the ground bending light waves would distort the entire image, including the background, in a wavy effect, not just one object. If everything around the object is clear, and only the object itself is blurred or distorted in a symmetrical manner, then the distortions are most likely coming from the object itself and not its surroundings.

CHAPTER 5

Exploration Opportunities Using SlipString Drive

With a ship of this sort, intergalactic travel would no longer take so much time as to be impractical as it is today. Possibly we could travel from one solar system to the next in a matter of minutes. This way, we would be able to really explore our galaxy and our universe in person. We would no longer be limited to robotic probes and the Hubble Space Telescope, as wonderful and beautiful as the Hubble's images and findings are.

Because of our ability to travel vast distances rapidly, a new form of research would now become available to us: Temporal Astronomy. Because light waves take 250,000 years to cross our entire galaxy and would travel much slower than we now could (in effect), we would be able to take advantage of this property. We could visit a solar system or other astronomical target of interest and then see what it looked like tens, hundreds, or even thousands of years earlier. This could be done by traveling farther away from the object, as the light waves emanating from the source are limited to C and still need to catch up to our ship. If a star goes supernova, and we did not have a telescope pointed at it at the time, we could watch it go supernova again and again. This could be accomplished by traveling a few light-days or light-hours farther away from the star and pointing our instruments directly at the object so we can take readings for the second or third time around and get all the measurements we needed.

If we did happen to find an intelligent civilization on another planet, by using highly sensitive instruments we could watch a hundred or more years of the civilization's progress up to its current state of development. This could be accomplished by flying farther away from the planet and watching their history unfold, as well as by replaying old TV or other radio wave or laser broadcasts from their past if they had that technology. This would open up all sorts of research opportunities, both astronomical and social.

We could also now protect ourselves from two of the most destructive astronomical phenomena in our galaxy: supernovae and the collisions of neutron stars. When a supernova occurs, it generates enough gamma radiation within a ten-thousand-light-year radius to destroy the chemistry of Earth's upper atmosphere, allowing ultraviolet radiation through at fifty times current levels. That is enough UV radiation to wipe out two-thirds or more of our planet's species (including all land animals), and there are enough massive stars within that radius to cause this sort of extinction every few hundred million years.[1] The last time a supernova occurred within this distance was probably about 440 million years ago, so we may be due for another one. Only some deep sea-dwelling creatures and species living deep underground in caves could survive this sort of disaster.

Another event that could wipe us out is when neutron stars collide. When they do, they release enough radiation to wipe out all life for about three thousand light-years for those systems caught in the twin beams of radiation emitted from the poles of the colliding system, and it would also do severe damage to systems even farther away.[2] If we calculated that a binary neutron star's orbit was unstable, or that a red giant was about to go supernova nearby, we could fly far enough away from the collision or explosion to be safe from it. Alternatively, we could hop in our SlipString Drive ships, create an isolated region of spacetime with our gravity waves, and be protected inside the region as the radiation wave passed completely around us. The radiation wave would pass harmlessly around our bubble, and we would be completely protected inside our ships. However, the planets nearby could be completely sterilized depending on the dose of radiation they received. We could reintroduce any life that we had protected on our ships onto the planet. This would be necessary in order to get the cycle of life and oxygen creation via photosynthesis back in motion, although nature would probably eventually accomplish this on its own. Be sure to save the plankton and bacteria! They are vitally important—we could not live on Earth without plankton which generates oxygen and bacteria which recycles dead organic material and purifies water. In fact, all the oxygen in our atmosphere today has been created by plant life on Earth, originally by plankton and later by trees as well. It was not until sufficient oxygen was created in our oceans and atmosphere that more complex life forms could evolve and breathe that oxygen. This gave them a greater energy source compared to the plants, and over time these animals were able to move onto land and eventually evolve into beings like us.

SlipString Drive could also give us advanced warning of impending threats such as those I just described. The farther we travel away from our solar system at speeds effectively faster than light, the sooner we would detect a supernova and the radiation wave it created heading toward Earth. Therefore, we could return to Earth and warn its inhabitants long before the radiation wave crossed the earth's path, giving humankind time to prepare for this type of disaster. The best way to survive this kind of mass extinction is to become a space-faring people, who are no longer tied down to a single moving target that cannot protect us from radiation such as this. A ship using SlipString Drive could not only avoid the radiation wave, but also by its very construction be very heavily shielded from radiation because mass is no longer much of a limiting factor as it is with current spaceship design. We could truly build cities in space, because by generating gravity wave bubbles around enormous objects, we negate their mass and would allow them to easily lift up into space and to effectively travel faster than light.

At this point, some physicists may think: "Hmm…but is not the transfer of information limited to the speed of light as well? Does this create a paradox?" Information cannot travel faster than C because it is moving through spacetime, and nothing can travel faster than C through spacetime. Physics does not currently take into account the possibility of anything bypassing large sections of spacetime. We could do this same trick using a wormhole between two points, and that would not violate physics. Using SlipString Drive, we are effectively bypassing long stretches of spacetime, being completely isolated from it, and there are no time-related issues to deal with. Yes, we can warn Earth of a radiation wave headed toward it before the wave hits Earth because none of that information ever traveled through spacetime faster than C. It was observed according to the laws of physics, entered SlipString Drive according to the laws of physics, exited, and was then delivered to Earth according to the laws of physics via radio waves, etc. No information ever actually passed through spacetime faster than C, and therefore we have no paradoxes to deal with.

The next disaster we could avert is that of mass extinction on Earth by an asteroid or comet impact. Small meteors strike the earth or burn up in our atmosphere all the time. However, about every 60–100 million years or so, a really big one gets through. An asteroid just a few kilometers across has more destructive power than all the nuclear weapons on our planet put together. One meteor wiped out the dinosaurs 65 million years ago and along with them the majority of life on Earth (although it may have been helped by a neutron star collision as well). It also happened more than 100 million years ago, and these extinctions may even occur on a fairly regular basis about every 62 million years, as indicated by our geological record.[3] Although there are still a few scientists who prefer volcanoes as the main method of extinction, a meteor strike would also set off a massive number of volcanoes due to its impact. Meteors and comets are the only way to explain a fairly regular pattern of extinction that happens when an Oort cloud object or our passage through the galactic plane disturbs comets and asteroids, sending them in our direction. This cycle of destruction and new animal evolution has happened many times over the earth's history, and it is just a matter of time before we get hit again. When we do, it will cause mass extinctions and possibly even our own demise if we do not invest in creating SlipString Drive ships (as described in the next chapter) in the near future in order to ensure the survival of our species.

When an asteroid of several kilometers in size strikes the earth, many bacteria will survive, but larger animals, including humans, will have a much more difficult time. The more food a species must consume in order to live, the lower the likelihood of its survival after so many of the species it depends on for food have died out. However, by using a ship with gravity wave propulsion, we could easily alter an asteroid or comet's orbit away from a collision course with Earth. By pushing the comet or meteor slightly with our ship (into another orbit that does not intersect Earth's), we could save it from disaster. There is no need to use clumsy nuclear weapons, which could easily split the comet or meteor into dozens of pieces. This could multiply its destructive power with a shotgun effect as each piece strikes different areas, areas that might have been spared the worst of the collision if the comet's orbit had not been altered.

Our moon was created as a result of the largest collision in Earth's history. Had the planetoid that hit us been any larger, or had it hit us dead-on instead of with a glancing blow, we probably would not even be here to discuss it. Our moon stabilizes our orbit and prevents the earth from tilting wildly on its axis, throwing weather patterns and temperatures into chaos. If the earth tilted randomly on its axis, stable agriculture would be nearly impossible. The planetoid that struck Earth also contributed most of its molten iron core to the earth's core in the collision because iron is so dense that it fell to the center of the collision. This left rocky mantle material in orbit to form our moon. Earth's spinning molten iron core generates a large magnetic field that protects life on Earth from solar and cosmic radiation. The molten core would cool down and solidify much sooner if we did not have a moon orbiting Earth and generating tidal forces. The tidal forces continually stretch out our planet so it is slightly wider than it is tall as the moon pulls on the earth with each orbit, generating heat in the core of our planet and the ocean's tides. This heat helps to keep the core molten and spinning. It also helps to keep the earth's magnetic field stronger for a much longer period of time than if no moon had been created in that collision. Without the moon, the earth's core may have already solidified by now, leaving us vulnerable to solar and interstellar radiation, which would have sterilized the earth's surface.

Tidal heating also helps to keep our carbon cycle intact. Plankton and plants process carbon dioxide from our atmosphere into oxygen, turning the carbon they receive into sugars in order to grow. After these plankton and plants are eaten by animals, the animals die, and some of them get trapped on the ocean floor or in tar pits on land, etc. The trapped organic matter, over millions of years, becomes coal, oil, natural gas, or other carbon compounds in our crust. Volcanoes then belch some of this trapped carbon dioxide back into the atmosphere, completing the cycle. However, not all of the carbon is returned to our atmosphere. Over time, more carbon gets trapped in the earth's crust, slowly reducing the amount of natural carbon dioxide in our atmosphere. Without this cycle, the carbon dioxide in our atmosphere would get completely locked into the earth's crust, depleting it from the atmosphere.

However, our current situation of massive carbon dioxide (CO_2) pollution from the emissions of vehicles and industrial plants throws this natural cycle out of balance. We cannot afford to keep doing this. The more that we pump carbon dioxide, methane, and other greenhouse gases into our atmosphere through pollution, the more the heat from the sun gets trapped by these compounds in our atmosphere, which causes our global temperature to rise. Volcanoes produce 110 million tons of CO_2 a year, but they can actually cool the earth by 0.1°C–0.5°C (0.22°F–1.1°F) for the first few years after they erupt. This is because of the ash they throw into the upper atmosphere, blocking sunlight and preventing it from reaching the planet. Human activities, however, produce 10 *billion* tons of CO_2 a year, the equivalent of seven volcanoes erupting every day, 365 days a year.[4] This has caused the amount of CO_2 in our atmosphere to rise nearly 30 percent over the last hundred years. Changing levels of atmospheric CO_2 correlate directly with changes in global temperature going as far back in our atmospheric record as we can, which is more than four hundred thousand years. Records from core samples of Antarctic ice going down several miles have confirmed this correlation between atmospheric CO_2 and global temperature.[5]

Climate change does not happen everywhere evenly. Some areas warm more than others, which causes global weather patterns to change. Some areas have droughts while others get much wetter as what used to be fairly regular weather patterns become more chaotic with the more energy that is pumped into the global weather system. Weather phenomena, such as hurricanes, are directly related to the ocean's temperature. The warmer the oceans, the more powerful the hurricanes that will be created, and the more money we will have to spend to clean up the devastation they create (not to mention the countless lives they end and suffering they cause).[6] As our glaciers (which reflect the sun's radiation, keeping the planet cooler) melt, which they are doing more and more rapidly, they expose more of the ocean and the earth's surface to the sun. The exposed oceans absorb the sun's radiation—as opposed to the glaciers which reflect it—and the warmer our planet gets in this vicious cycle or feedback loop.[7] While the average global temperature has risen "only" 1° F so far, Arctic temperatures have risen 8° F, largely because of this feedback loop, causing massive glacier melting. Sections of the pole the size of Manhattan Island are regularly breaking off and floating out to sea as they melt, which also raises the sea level.

Venus became a super-hot, acidic, volcanic wasteland because of a runaway greenhouse effect. Venus started out very much like Earth, but because Venus is closer to the sun, it absorbed much more of the sun's radiation. Venus heated up so much that its oceans started to evaporate. Water vapor traps heat even more efficiently than carbon dioxide does, so once that much water vapor was in the atmosphere, Venus was doomed. So much heat was trapped by its atmosphere that the oceans completely boiled off, and the water vapor was eventually split into hydrogen and oxygen atoms. The oxygen rusted the surface as it combined with iron, and the hydrogen combined with sulfur atoms to create sulfuric acid which is now a main component of Venus's deadly clouds.

Global warming may or may not get pushed to this extreme here on Earth, but it's a sobering fact that we do not know what the trigger temperature is to cause this catastrophic scenario until we hit it. In the meantime, the earth's weather will continue to get more violent and unpredictable. Our sun will continue to get hotter as it ages (about 10 percent warmer every 1.1 billion years) because it uses up more of its hydrogen fuel and starts to fuse more helium and heavier elements together, producing more energy.[8] The early Earth was very geologically active after the impact that created our moon, throwing huge amounts of greenhouse gases into our atmosphere. That was not a problem at the time because our sun was radiating about 30–40 percent less energy, and all those greenhouse gases helped to keep the earth roughly as warm as it has been over the past several millennia.

As the earth aged, it continued to cool down and became less geologically active. It was also struck by fewer meteors which could also increase geologic activity. In addition, ever since life has evolved on Earth, our planet has been removing carbon dioxide from our atmosphere and "trapping" it inside the earth in the form of coal, oil, methane, and other carbon-rich compounds. Only part of this stored carbon is reemitted from the earth via volcanoes and other geologic activity. The rest stays trapped, removing more CO_2 from our atmosphere as the earth ages, which keeps the planet's temperature relatively constant as our sun continues to heat up. However, if we continue to dig up and burn the earth's entire store of carbon compounds, releasing all those greenhouse gases back into our atmosphere while at the same time our sun is continuing to increase its temperature, we are asking for trouble. This is, unfortunately, just one more reason that we will need to be able to leave our planet on a moment's notice and to construct SlipString Drive ships to preserve our species if the worst happens.

Another use for these ships is for deep-sea exploration. With gravity waves powerful enough to force spacetime apart, we could easily resist the pressures of the ocean down to its utmost depth. We could even travel at supersonic speeds underwater as our symmetrical gravity waves would repel the water around our ship, creating zero drag. We could also pass through ice by using a combination of repelling spacetime around the ship (including the ice) and by melting a narrow tunnel through the rest of the ice with the heat produced by our engines. We would be able to explore every last square inch of the ocean and discover all sorts of new species. We could explore deep-sea volcanoes and observe all the species that live off the chemicals being spewed out of deep-sea thermal vents. With SlipString Drive, we could easily explore the frozen moons of gas giants far out in our (and possibly other) solar system, such as on Jupiter's moon, Europa. Europa probably has a huge liquid water ocean beneath its crust of solid ice and the best chance for other complex life forms existing in our own solar system.

As strange as it sounds, we could also use SlipString Drive technology to explore the human body. Just before publication I realized that the gravity waves produced by the engine of a SlipString ship could also be used in small amounts to perform delicate surgery. A small probe that redirects a few gravity waves from a ship's engine could push tissue away from the probe without damaging a patient's tissue or even leaving a scar! Gravity wave surgical instruments could perform delicate operations by curving skin, muscle, bone, and other tissues away from the probe, allowing access to organs or other areas that require repair. As with spacetime, the tissue that was moved by the gravity waves would be completely intact, and circulation would flow normally around the curved space generated by the probe. Once a space was opened up, surgery could be performed or medical implants could be inserted such as medication pumps, pacemakers, and so on. Once a small pump or pacemaker was inserted and the probe was withdrawn, the muscle and other tissue would simply lay back down on top of the new device without being disturbed or traumatized as it would be with a traditional scalpel.

Technology of this kind could revolutionize the practice of medicine. Similar to laparoscopic surgery, but even less invasive, gravity wave surgery would make nearly any type of surgery safer, nearly painless, and much less traumatic for the patient. It would also be much more forgiving for the surgeon, who no longer needs to spend lots of time cutting through tissue in order to access the region they need to work on. Surgery near the spinal cord is extremely dangerous as bone must often be cut away using jarring techniques and one slip could paralyze you for life. Using gravity waves, there would be no incisions, cutting, or other harmful techniques necessary to access the area around the spinal cord or any other delicate part of the body for that matter. For these reasons alone, investment in gravity wave technology should be a top priority. Generating the relatively small amount of gravity waves necessary for this application should be much easier than generating those necessary to propel a ship. This kind of technology could be possible within 20 years, given enough investment. In addition, there are probably many more beneficial applications of gravity wave technology, medical and otherwise, not yet considered in this chapter.

CHAPTER 6

Gravity Waves and How to Manipulate Them

How to generate gravity waves—a pretty tough feat, but the benefits to humankind could be enormous. Gravity waves that we might be able to detect in the not-too-distant future are created in violent events such as supernovae and binary black holes that spiral in toward each other before one black hole gobbles up the other. These waves can travel across the universe, although detecting them over great distances becomes very difficult because these waves, as with gravity, get weaker according to the inverse square law. This means that for every unit of distance the wave travels, it gets weaker by GM/r^2 ("r" being the radius from the gravitational source). So if a gravity wave travels one hundred light-years, the wave will only be 0.0001 times as strong as it was just one light-year from the source. So instead of compressing spacetime by ten meters one light-year from the source, it would compress spacetime only by 0.1cm one hundred light-years away. As large gravity waves rarely originate so close to us, we are usually talking about such tiny units of measure that they are extremely difficult to detect. Unfortunately, we do not have a way to detect these waves on Earth yet because our own planet's gravitational field obscures our view of other gravitational forces, but several teams have begun the effort. You can help by volunteering your computer's free time at Einstein@home by visiting http://einstein.phys.uwm.edu/.

In the next few years we may be able to detect these waves by putting lasers in orbit far away from the earth's gravitational field to measure ripples in spacetime. To do this, we will first need to put three mirrors (or detectors) in space, each one on a different axis (x, y, z) with a laser emitter/detector at the origin of the three axes. Each mirror needs to be exactly the same distance (several miles) from the laser/detector at the origin and exactly ninety degrees apart from each other. The laser/detector at the origin of the three mirrors measures exactly how far away each mirror is (see fig. 6.1). Then, if a gravity wave passes through the detector, the distances between the detector and the mirrors will shrink briefly and then expand back out to their original positions.

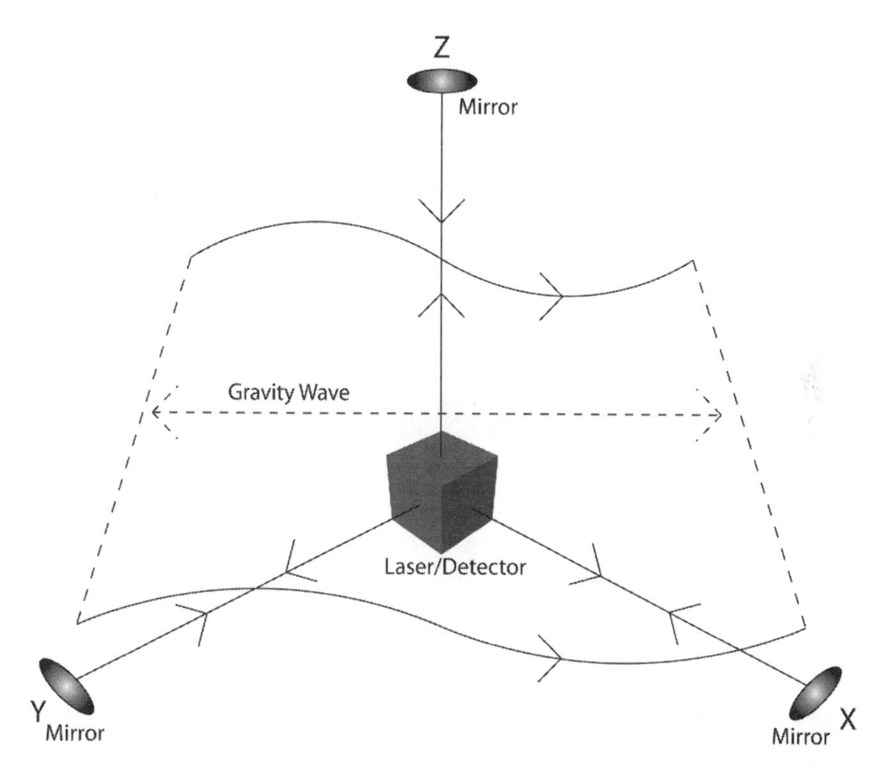

Fig. 6.1. *Detecting gravity waves using space-based detectors. Each mirror is exactly the same precise distance from the laser emitter/detector which would bounce a laser beam off each mirror, measuring its distance from the detector. Gravity waves are detected as the distances between the mirrors and detector shrink slightly and then expand back out to where they started. With two sets of detectors separated by a large distance, such as the orbit of Earth or Mars, we would then have stereoscopic vision and be able to determine the exact direction and distance of the source of the waves. Then, we could observe it with optical, infrared, X-ray, and other telescopes to determine the source of the wave.*

By calculating the exact distance that each mirror was pulled toward the detector and which axis was affected first, second, and third, we could then calculate the strength of the wave and the direction from which it originated. If we have two sets of detectors separated by the diameter of Earth's or Mars's orbit, then we would have stereoscopic vision and could pinpoint exactly where the wave originated. Then we could follow up with optical, infrared, X-ray, and other telescopes in order to see the source of the gravity waves we just detected.

Producing highly concentrated gravity waves is going to be a very difficult feat. Gravity waves in nature are weak by the time they reach us, but can be exceedingly strong near their source. We will need to spend quite a bit of time, energy, and resources in order to develop a compact way of producing and guiding our own gravity waves in order to build the ships necessary to explore our universe, and save humanity from a dying planet or other disaster.

Once we did manage to create a source capable of producing sufficient gravity waves, we would need to figure out how to control those waves in order to guide them evenly around our ship. We would also need to figure out how to guide additional waves below the ship for maneuvering near planets, and also aim waves behind and in front of the ship to stretch spacetime for "faster than light" travel. Additionally, diverting enough waves to simulate at least half of earth's gravity in the crew cabin would help to keep the crew healthy. We would need to divert the same number of waves from above and below the cabin in order to make sure our gravity waves are precisely matched so that they don't affect our maneuverability. This simulation of at least half earth's gravity would help to prevent the loss of bone mass and the weakening of the heart and other muscles which currently are problems of extended exposure to the weightlessness of space that our astronauts currently must face.

In order to focus any gravity waves we might create, we could take advantage of wave properties similar to those of light. Light waves slow down as they pass through denser materials such as glass or diamond, which can be shaped in order to focus light waves. Gravity waves should similarly travel slower through more massive elements than less massive ones as they actually have to travel a longer distance through the more massive elements than they do through less massive ones. This is because heavier elements curve spacetime more than lighter elements due to their increased mass, creating a longer pathway for the gravity waves to travel through. We could use this effect similar to a traditional lens to focus these waves or to create differences in speed to bend the waves around curves in our waveguides. By using a variety of elements to achieve the degree of curvature we desire, we can cause the gravity waves to pass from one element to the next, altering their paths as they do.

Once our technology becomes advanced enough, we could eventually produce liquid lenses capable of focusing gravity waves extremely rapidly and precisely. A computer controlled liquid lens made of a mercury-like liquid metal compound could be curved by passing a current through it, causing the liquid to change shape, and hold it, changing the wave's path where we need it to go at any given fraction of a second.

Although creating gravity waves may seem like a pipe dream to some, I have confidence that we will find a way within the next 50 years or so. Now that we have the goal, and a major reason for attempting to produce them, it shouldn't be too long before we find a practical way of producing such waves.

When I started writing this book, I thought that due to the method of travel proposed, attempting to start a war with another species using SlipString technology would be pointless as either side could simply "disappear" at will, avoiding any damage. While I still believe this, that doesn't mean the technology is completely without possible harmful uses. Like any technology, it can be used to benefit or harm society. If used on earth, for example, (instead of simply for space exploration,) individual rights could be abused. The governments which possess such technology could use it to "break and enter" on someone's private property without even leaving a trace of having been there. They could literally exit a craft within someone's house without disturbing the neighbors. Hopefully, now that we realize there is a possible danger here, laws will be passed in the near future in order to limit the use of such technology to space-based applications only, and not for personal surveillance. It would be much too easy to abuse the technology without strict governmental limits allowing its use only in the most dire of circumstances.

Our technology and understanding of our universe are beginning to reach a point where we will soon have such advanced technology and understanding of physics that we could end up damaging our entire multiverse if we aren't exceedingly careful. Some scientists have discussed their desire to create a universe in their lab. According to string theory, it shouldn't be all that difficult to do, either. Such a universe would simply splice itself off of ours, and grow out in our multiverse. The only problem with this is that one never knows what will evolve in such a universe, or whether the universe itself or what evolves within it will be able to come back to haunt us at some point. It's usually best not to play God where universes are concerned. You never know what you're going to get.

With a complete understanding of physics and our universe, fleets of powerful gravity wave ships and quantum computers (which, in the future, could accurately predict upcoming events with high degrees of probability among other things), humanity will need to be exceedingly careful as to how it will proceed. We are developing quantum computers which could, down the road, allow us to manipulate events, or these computers could even think on their own (which is a *very* bad idea if I do say so myself). Given the right conditions, a quantum computer could become self-aware, and with the ability to "look ahead" at future simulated events, and even simulating people's thoughts, the computer itself may be deciding your own fate instead of yourself!

With the combination of quantum computing and nearly complete knowledge of our universe we could gain the power to manipulate not only our but even other universes as well. While these warnings may sound quite overblown at the moment, these powers will become so extreme that in the wrong hands, that we could destroy everything we hold dear from our Constitution to free will. Power could be consolidated so rapidly that if we are not eternally vigilant, those in power could stay in power ad nausium. This may begin with our rights as citizens being chipped away at, such as with a wiretap bill that allows *anyone who receives email* to be tapped, or a Patriot Act which no Congressperson reads, yet allows for a US citizen who never broke the law to be killed if deemed a threat by only one person. These beginnings could easily evolve into something that threatens our and possibly even other universes if we're not exceedingly careful. That is, if we discover a way to travel between universes. (This may seem like a pipe dream, however, with string theory and Membrane Gravity, it might actually be possible to travel between universes, and that may not be a good thing.) We have plenty of real estate right here! Why bother the neighbors?

I know many Trekkers would love to have their own personal Data android, but if we actually gave an android or a ship its own learning quantum computer, we will have no control of that technology if it gets out of hand, or if we lose control. Regular computing technology is sufficient to do nearly everything we need at the moment, and possibly in the distant future, including building robots or navigating ships. Quantum computers may sound "cool," but they are completely different devices altogether. If we designed one that could learn on its own, and it was completely mobile like an android or SlipString probe, we would essentially be creating something which we will have no control over *when* it gets loose. I wouldn't want to be around during its childhood!

Quantum computers need to be limited to calculating models or scenarios that *we program into them*, or programmed for specific tasks such as calculating the curvature of light around a SlipString ship as it maneuvers so a pilot can see clearly. This may be more time consuming than creating a learning quantum computer, but if we are to go down the quantum computing path, it is a very necessary limitation.

If a quantum computing SlipString probe was given the ability to learn on its own, it could essentially become a child with the power to make enormous mistakes whose consequences could literally destroy civilization as we know it. Such a probe could learn to reproduce itself, and once done, you have a fleet of curious probes looking to learn in ways you never could have imagined. They would have the power to move planets, manipulate spacetime, and otherwise create havoc, yet lack the wisdom to use (or not to use) those powers responsibly.

OK. Now you might be wondering, "If he thinks this stuff is so dangerous, why the heck did he publish this darn book in the first place?" Well, this book isn't a "how to" manual. There are few if any equations to help designers construct anything for one. Secondly, it's only quantum computing that makes all these scenarios possible, and I have no power over that, which is why I'm writing this warning. Lastly, science will always progress. If it wasn't me telling you that SlipString Drive was possible, twenty or fifty years from now, someone else would. At least this way, you (and hopefully the scientific community) have been advised well ahead of the game, and will look out for any warning signs, and with any luck prevent the integration of quantum computers into a mobile package. Hopefully, those who design quantum computers will be smart enough to keep them limited in their capabilities even when they are able to be "enhanced" with high technology. Computing is enough. There is no need whatsoever to give a quantum computer the ability to move freely and learn at the same time.

Chapter 8 deals with our entire universe from before it started until after it ends. It has been a goal of humanity to figure out how our universe works ever since the first of us peered up at the stars millennia ago. If I hadn't also "put the icing on the cake" of M-Theory, again, someone else most likely would have down the road. At least this way, you have my warnings as well. Hopefully, most of us will be happy to know how our universe works visually, but won't need to derive all the equations for ourselves.

In order for my warnings to come true, all three fields: quantum computing, SlipString Drive, and complete knowledge of our universe will need to be united. Somewhat like our three branches of government in the US, I am hoping that we are wise enough to stay gridlocked, and never let all three branches of power fall into the same hands at the same time.

We face a wonderful yet scary period ahead. We have amazing technologies which could make our lives easier and less painful, or much more complicated and scary. I believe we have the ability to get past this era unscathed, but it's going to take a huge amount of willpower on our behalf, and hopefully we will all evolve to a point where greed and anger no longer control us. We must if we are to succeed.

CHAPTER 7

SETI (Search for Extraterrestrial Intelligence) and How to Fix It

In recent years, the search for extraterrestrial intelligence has really taken off, thanks in large part to the launch of SETI@home by The Planetary Society at the University of California at Berkeley. I have been a member of this wonderful organization for more than a decade. The search for signals from "E.T." now has a massive amount of computing power behind it, thanks to millions of average computer users and a much better chance to succeed due to this increased power. I personally have processed well over seven thousand SETI work-units, as have thousands of other computer users. The thought behind SETI is that because space is so vast it is unlikely that other technologically advanced species will be able to travel to our planet. Therefore, they will have to resort to trying to locate each other by sending out powerful radio waves that will travel for at least a few hundred or possibly even a few thousand light-years (with better equipment). The current programs scan the sky for strong radio (and soon optical) signals, and once a signal is detected, other telescopes are aimed at its location in order to verify that it is genuine.

Here is the problem with the current method of verification: Whenever a good signal is detected, it takes other astronomers several minutes to an hour to swing their radio telescopes at that location, by which time the signal is gone. There are several good reasons why this has been our experience. First of all, 75 percent of all the solar systems in our galaxy are older than ours. This means that, most likely, 75 percent of all technologically advanced alien civilizations are much older than ours and, therefore, much more advanced. These civilizations have probably figured out how to travel using SlipString Drive techniques (or something similar). Thus, they can travel easily throughout our galaxy and other galaxies and have no need for huge, unfocused radio-wave generators to locate or communicate with other advanced species. They can simply fly around and investigate other solar systems on their own and interact in person if necessary.

They will, however, need to communicate with each other and with other species that they come across in their travels, as well as with their own probes and satellites, etc. Over long distances, it would be much more efficient to send out SlipString probes with data recorded inside them than to use radio or laser transmission. Unlike the probes described in chapter 6, they would have "regular" computers (not quantum) with strict rules so there's no chance of them becoming dangerous. Actually, data may need to be sent in person, pilot-to-pilot, so we keep the personal touch alive and don't become too reliant on probes which may get out of hand. Probe-delivered communications will not be as slow as radio signals, which are still limited to the speed of light and take many, many years to reach their destination. These probes (or ships) could be filled with whatever data needs to be sent and would travel the same way SlipString ships do. Unfortunately, this would be completely undetectable to us or anyone else, as they would disappear from view when entering their isolated regions of spacetime. However, over shorter distances (such as within a solar system) radio and light waves are still an efficient means of communication. So it is still possible that older alien civilizations could use highly focused radio or light waves for short-range types of communication only. Other younger civilizations (such as ours) could still communicate using radio waves, but they would only do so for the brief period of time between developing radio technology and developing SlipString Drive technology. This is a very small window of time, probably just a few hundred years, which is nothing compared to the time it takes for a species to evolve.

We could easily intercept some of these short-range communications (and probably already have), but by the time we turn another radio dish to track the signal, it will already be gone. In order to develop SETI to look for these sorts of short-duration signals, what we would need to do is record as much information about the signals as possible. Preferably, we would need to record the entire signal in as much detail as possible for analysis (which may be many gigabytes of information per signal). We could then send out the signal's vital statistics to SETI@home users to determine if the signal is of extraterrestrial origin and, if not, delete the recording in order to save storage space. All of the analyzed signals that are not ruled out should be scrutinized in detail to determine if they have any intelligent structure to them. If so, we should get to work analyzing the signals and attempting to decode them.

This way, we would finally be able to determine if alien species really are still using radio or optical signals to communicate and what type of information they are sending in this manner. The scenario in the movie *Contact* is probably not the way alien species work (even if it was a wonderful movie). Why would alien species send out signals in such a way that could be misinterpreted by less advanced (younger) species and possibly injure the young species' fragile, self-centered view of the cosmos? (Not to mention interfering with their natural development.) Alien species with SlipString Drive can simply fly over and check on another planet's progress without contaminating their culture very much. If a few people claim to have seen flying discs or strange lights in the sky, they can always be labeled as crackpots by the culture at large and ignored for the most part.

Imagine that in the 1930s someone discovered an alien message via a radio transmission. It is highly unlikely that the culture at large would have been ready for alien contact that long ago, and humans certainly would not have had the technology to act on the signal, such as building the construction in *Contact*. Better that these early technological civilizations were blissfully unaware of any extraterrestrial existence, so that they will go on with their lives and discover other planets and eventually alien species with their own ingenuity. It would be best to let younger civilizations prove their worth as a species before handing over some dangerous alien technology they are too immature to handle.

Once a species learns how to use gravity waves to curve spacetime enough to isolate a region and to propel themselves quickly throughout the universe, then they are worthy of being dealt with as a partner. Before then it is too dangerous to simply hand over some advanced technology and hope that they will use it correctly. *Star Trek* had this much correct. The "prime directive" of noninterference with non-Warp Drive-capable cultures (or in this case SlipString Drive-capable species) not only protects the advanced species but also prevents the younger species from annihilating themselves with the technology before they are ready to handle it on their own.

By the time cultures perfect SlipString Drive, they will have already imagined themselves exploring the stars and the possibility of extraterrestrial life around those stars. At that point they will discover other solar systems on their own. The amount of time between discovering radio waves and developing an efficient form of space travel (perhaps one hundred fifty years) is the blink of an eye in the time frame of the universe—and probably for most space-faring species with thousand-year life spans. Not to mention that during those one hundred fifty years the young civilization will have learned to split the atom and create nuclear weapons. If they wipe themselves out with those weapons, they are probably much too dangerous to deal with anyway.

Returning to trading with other species, this will create another major danger that we may soon face. The moment that we do develop or "repair" SlipString technology, we will be tempted to trade with other species (if they exist) for all sorts of advanced technology. While this may sound "cool," I have serious doubts as to the readiness of our current government (and even humanity itself) as to their ability to use all these technologies responsibly.

For example, I wrote a blog entry about the connection I felt with my cat companion of 22 years, and her connection with me. She always knew when I was coming home, even before my vehicle was visible, and would look out the window anticipating my arrival well before I did. We were very close, and she was very protective of me. When she eventually did die, I immediately felt like there was a massive decrease of blood flow to the right frontal lobe of my brain. This feeling continued for several months as it gradually faded. This may have been an example of a very weak form of quantum teleportation at work.

Studies have been done at Princeton by Physicist York Dobyns at their Engineering Anomalies Laboratory demonstrating quantum mechanical "spooky action at a distance," such as quantum teleportation, albeit very weakly.[1] They demonstrated that if someone concentrates hard enough on, for example, a computer program flipping coins, instead of the results getting closer to 50/50 the more flips of a coin made (as it should), by concentrating on "heads" or "tails," a person could influence the results very slightly.

Since the mammalian brain is so much more similar to one another than to a computer system, it is possible that a very weak form of telepathy could have evolved thus far. There are crystal structures in the brain which may be possible of creating/splitting photons for this function. Both brains would observe the spin of split photons simultaneously, causing the wave function to collapse, and communicating the quantum state of one party's nerve signals to the other.

If this is true and weak forms of telepathy are or may be possible, it is likely one of those types of technology too hard for certain government officials to resist. If the technology is traded for, and not created by our own ingenuity over several decades to hundreds of years, we will likely abuse the technology, having no laws to govern it. It would be too easy to observe the minds of those who never had a criminal record, but simply hold views that an administration simply doesn't agree with. If we ever do reach the point of telepathic communication through technology, it would need to be the most tightly regulated and least used tool in our toolkit. We are approaching the most dangerous time in our species' history. We must learn to use our new technologies wisely and with an incredible amount of restraint. If we don't, we may see our society's freedoms fall one by one.

In general, once a species moves out into space, it will probably start tinkering with its own genetic code in order to make the experience a more pleasant and convenient one. It would not be surprising if other advanced species (if they exist) have engineered themselves new organs to turn electrical energy into proteins, sugars, or other ways to maintain themselves. They could use some sort of "electrosynthesis" to absorb and convert excess energy directly from their ship. Carrying tons of food for hundreds of thousands of light-years is not very practical. Instead, you could periodically replenish your supply of fuel in order to feed your fusion or antimatter engine, giving yourself a huge amount of energy at your disposal. It would make life so much easier to use that energy for the majority of your maintenance rather than having to restock huge supplies of rations and constantly deal with waste removal and purification systems. However, in the short term, that is what we will have to deal with.

Currently, when humans spend any extended amount of time in a weightless environment, our bones start getting weaker, as do all the muscles including the heart. We will need to rewrite our genetic code to constantly maintain our bone structure and create muscles and bones that do not atrophy when not exposed to gravity. It may be easier, however, to simply divert a few gravity waves inside our ships to simulate about half of Earth's gravity to reduce most of these problems. We will also need to extend our life spans into the hundreds or even thousands of years. It would be impractical to continue reproducing every twenty to thirty years.

Recent genetic modification of animals, such as mice and fruit flies, doubled their life spans and gave them a higher quality of life by simply tinkering with a single gene.[2] Other studies have extended mouse life spans by 50 percent by genetically increasing a single protein produced by the mouse.[3] With these and other approaches, I would not be surprised if we could double and eventually quadruple human life spans over the next few hundred years. In combination with advanced future modifications, such as "electrosynthesis" and rewriting a few organs for efficiency, life spans of up to one thousand years would not surprise me. Current life spans are relatively short, because evolution requires that we produce new generations every twenty to thirty years in case a natural disaster strikes so that we have the genetic diversity for some of us to survive it.

We may need to take extended journeys of possibly hundreds or even thousands of years in order to leave a dying galaxy or universe. Eventually, Earth will become a lifeless rock. It may be caused by the earth's core cooling down to the point where the carbon cycle can no longer continue. Or, it could result from a runaway greenhouse effect caused by our own pollution and shortsightedness, turning the earth into a toxic, boiling, Venus-like planet. Our sun will eventually go nova in about five billion years, leaving the earth a burnt cinder. At least by that point (but probably much sooner) we will need to leave this planet and take our place among the stars.

Our nearest neighbor, the Andromeda galaxy, is accelerating toward our Milky Way galaxy, caught in its gravitational pull. These two spiral galaxies will collide and merge in approximately three billion years.[4] This could send millions of solar systems, including ours if it is still around, hurtling out of the Milky Way galaxy and into empty space in a long chain of stars. Another possibility is that our solar system might have the bad luck of being on the colliding side of our galaxy when the two merge, sending our system straight into the chaos of the collision—toward the enormous black hole at the center of Andromeda! The collision of galaxies is likely to cause severe damage to our solar system if it slingshots around another star or black hole. Such a near miss would jettison it away from our galaxy and into the blackness of space. We could end up far away from any other stars, perhaps even our own.

We will need to leave our galaxy during this process and possibly follow the long chain of solar systems that will be thrown clear of the collision. We could travel out into deep space until our new combined galaxy settles down. The two spiral galaxies will become either a new elliptical galaxy or a larger, more chaotic spiral galaxy. The black holes that were in the center of each galaxy will circle one another, spiraling in, and eventually merging into a single, even more massive black hole. By this time, we could return to our newly reorganized galaxy for several billion more years and explore the results of the collision and new star formation created by the collision. After several hundred billion years as our universe continues to expand—pushing galaxies farther and farther apart—we will be forced to leave this universe altogether. The last of the stars will begin to burn out; expending all their fuel, and this universe will start to become a dark, empty place, as we will see in the next chapter.

CHAPTER 8

How to Complete a "Theory of Everything" by Modifying M-theory with a Membrane Theory of Gravity

From my perspective, over the last century many cosmologists have taken their theories of the universe too far (to the infinite degree) to be plausible today with the latest developments in theoretical physics and in string theory. For example, take theories of black holes. It was believed by most physicists that after the supernova of a very large star, a huge amount of matter was compressed into such a tiny volume that the star's remnants collapsed into a "singularity" at the heart of a black hole. A singularity is a one-dimensional point which most twentieth-century theorists believed was created if enough matter collapsed into a tiny volume. However, seen through the lens of string theory, the core of this black hole probably does *not* collapse into a singularity. Instead, matter gets compressed into a very small core just a few tens of meters across, forming the core of the black hole. Normally, the distance between quarks in a proton is extremely large compared to the size of the string that makes up each quark, due to the forces generated by each vibrating string.

When a smaller star dies, it does not have enough mass to create a black hole. Like our sun, when a star can no longer generate energy by fusing heavier and heavier elements together in its core, it will collapse into a neutron star. This neutron star is only a few miles across because there are no electrons orbiting the individual protons and neutrons within the star, but are instead orbiting the star itself, similar to an atomic nucleus. This allows the matter in the star to become compacted into an incredibly dense sphere—so dense that a cubic inch of its material would weigh thousands of tons—that still has the mass of a star like our sun.

As opposed to a neutron star, a black hole has so much mass that it would overcome even the nuclear forces that hold the three quarks (that make up each proton and neutron in the neutron star) apart from each other. This would mean that the center of each quark (a single string) would no longer have the strength to overcome the force of gravity. The field that makes up the quark's shell (similar to electrons orbiting the nucleus of an atom) would collapse. This would cause each quark's central string to get compressed right up against its neighbors with hardly any space in between them.

Therefore, a Type I black hole (the mass of about twenty suns to possibly more than thirty-five suns) could have a core only several tens of meters across made of highly compressed individual strings. (Again, strings are the Planck length of 10^{-33}cm or 10^{-35}m each.) This would curve spacetime so much that it would create an "event horizon" far away from the tiny core of strings inside the black hole. The compressed bundle of strings at the core would be so tightly compacted that they could no longer vibrate independently of each other. This would cause the strings at the core to act as if they were a single elementary particle, because each black hole is left with only three defining properties: mass, force charge, and spin (similar to a single electron).[1] The center of a black hole would no longer be a one-dimensional singularity (or a tear in spacetime) as was originally theorized. The core of the black hole is merely so much smaller than the enormous gravitational forces it produces. The event horizon of the black hole would continue to grow even farther away from its core as the black hole swallows more matter (because the core's size grows so little relative to the gravitational effects produced by the matter falling into the hole). I developed this hypothesis in 2003, and in 2004 other string theorists developed theories along the same lines. They proposed that information is not destroyed within a black hole, as it would be within a singularity, but is recycled as Hawking radiation with the same information as the matter that was swallowed by the hole.[2] During the last two years, most physicists have come around to this point of view; that information is recycled from black holes as Hawking radiation, eliminating the singularity and assuring that information is not lost within a black hole.

The same principle just applied to black holes applies to the big bang as well. If we run the universe backward from its current expansion to the big bang itself, previous theorists simply went too far back. They assumed that the universe started from a single one-dimensional point before it inflated. In string theory, theorists have attempted to go back to a Planck length-sized ball, which is more plausible than a singularity, but there are still many problems with theories attempting to link string theory (or any theory for that matter) to a universe expanding from such a small size. Currently nobody has been completely successful in trying to model this process. However, M-theory has come much closer with its proposal for a collision of membranes to create a new universe. This is very promising, but as yet it is not exactly clear what mechanism transfers energy from the collision into the strings of the new universe. I hope to clarify this process with a *membrane theory of gravity*.

Before string theory the standard model for the big bang starts off with an infinitesimal point with nearly infinite energy. This point of incredible heat and energy was thought to rapidly expand into a hot soup of subatomic particles, which, as it continued to rapidly expand, began to cool down. Four hundred thousand years after the big bang, the universe cooled down enough so that the subatomic particles bouncing around in the initial quark soup condensed into protons, neutrons, electrons, and antimatter. The protons and neutrons were attracted by their nuclear forces and gained an electron or two forming hydrogen, helium, and deuterium atoms (a hydrogen atom with an extra neutron). The antimatter created in the bang was instantly annihilated when it came in contact with normal matter. This released pure energy in the form of gamma radiation. After 13.7 billion years of the expansion of the universe, the our universe cooled to the point where a flash of visible light was emitted as the quark soup created in the bang underwent a phase change into the protons, neutrons and electrons we are made of today. This occurred about four hundred thousand years after our universe had expanded and we now see it as the weak cosmic microwave background radiation observed by the COBE and WMAP satellites.

Most string theorists now believe that the strings comprising our universe were created in a collision of membranes, as described in chapter 2, and that our universe is an expanding bubble of strings with nothing beyond it on our membrane. I propose, however, that it was the virtual strings that existed within our membrane at the instant of the "big splash" (when our membrane collided with another one creating our universe) that were vibrated into strings of matter and antimatter when our universe was created. According to Brian Greene,[3] Michio Kaku,[4] and a number of other physicists, even nothing is highly unstable on a quantum level. Due to this quantum noise (see fig. 8.1) on microscopic scales (sizes at or below the Planck length), tiny bubbles start to form. These bubbles (strings) vibrate in eleven dimensions (one of which is time), and when they are given energy from an external source, they will vibrate in ways that make them take on the properties of matter and energy. When virtual strings are generated, it is as if a quantum disturbance shakes waves of spacetime apart briefly. When this happens, one could observe the pair of virtual strings orbiting each other before they join back together and vanish as their waves cancel each other out again. Virtual strings are constantly being disturbed into existence, orbiting each other, and then vanishing again in what was thought to be empty space due to quantum noise. They are the basis of Hawking radiation, theorized to be emitted from black holes when one of the virtual strings falls into the hole, and the other one escapes and is accelerated away from the hole and out into space.

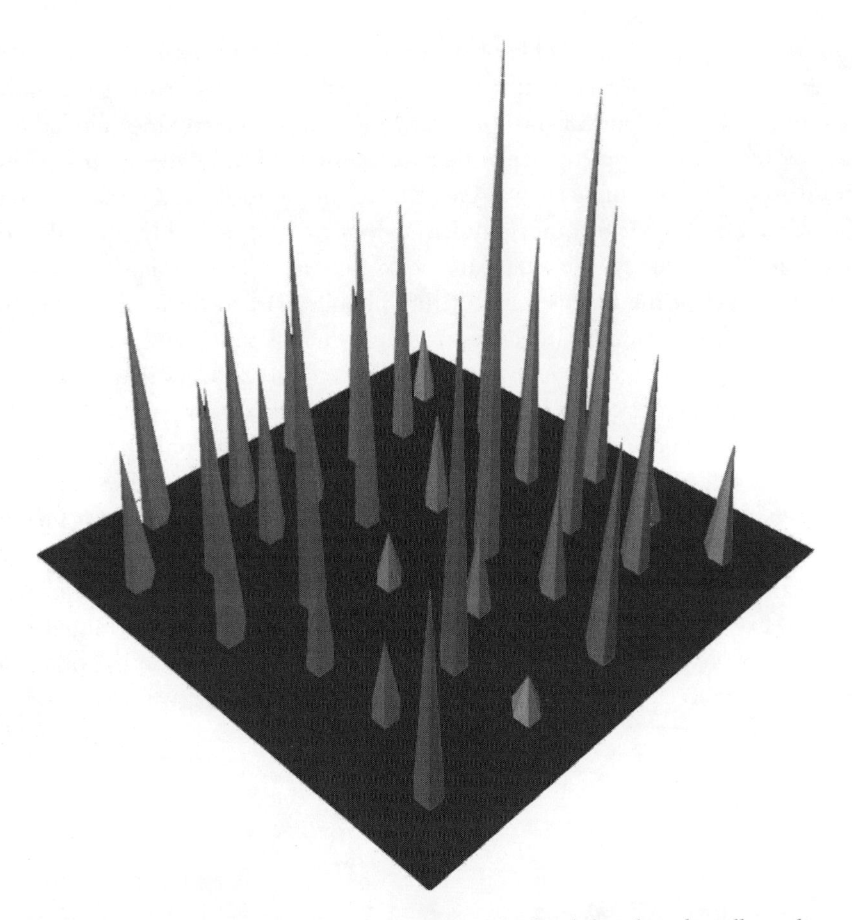

Fig. 8.1. *Computer models of quantum noise. On Planck-length and smaller scales, this noise appears as tiny bouncing spikes of static, similar to a randomized graphic equalizer on a stereo bouncing to radio static. This noise agitates spacetime on tiny Planck-length scales, forming tiny vibrating strings of spacetime. These strings can inflate into huge membranes, from which other virtual strings can then be separated whose ends are connected to the brane. Eventually, huge 4-branes (with 3 spatial and one time dimensions) full of virtual strings generated from this quantum noise gain momentum as more and more strings are created and destroyed within them and interact briefly with each other. With all this activity, it is likely a matter of time before two of these large membranes will collide with each other, creating new universes.*

If M-theory is largely correct, then what came before our universe could have been these tiny multidimensional strings or bubbles of energy being generated from the quantum background noise, such as Kaku has postulated on *Horizon,* the BBC TV series. Similar to how a breeze can cause a soap film to generate bubbles, and bubbles within bubbles, these vibrating virtual strings (and membranes) could be generated. In the process, branes and strings smooth out spacetime from what were rough, choppy spikes of quantum noise. Over billions or trillions of years, more and more membranes and virtual strings could be agitated into existence because of this instability. The branes could then stretch, inflate, and interact with each other, creating a much more complex and dynamic system. Eventually, these membranes and the virtual strings within them could fill up huge volumes of space that might be billions, trillions, or more times larger than our own universe.

Some of the first membranes to evolve could expand into huge sheets, bubbles, and many other multidimensional shapes like a torus (the shape of a doughnut) when they inflate from the vacuum around them (which has negative pressure, like suction).[5] Then, more virtual strings could be disturbed into existence within these branes, possibly inflating them further. Virtual strings generated within these branes have the potential to become matter and energy if they are vibrated rapidly enough in a big splash collision of branes, as will be explained in a moment.

The interactions of virtual strings within each brane create complex motion within the branes as they expand. As a brane's virtual strings interact with each other and the brane, the surface of the brane could start to bend and ripple similar to waves on an ocean. Branes could also gain momentum as they expand and the number of virtual strings being created and destroyed within them increases. With all this motion and activity, two nearby branes will eventually collide with one another. Subsequently, a multitude of branes would be generated and float around in the eleventh dimension, generated from the quantum noise that pervades the multiverse and colliding with greater force and frequency as their numbers increased. Currently, M-theory does not specify exactly how the energy of a membrane collision is transferred to the new universe within each brane, which has made it difficult for physicists to take the theory seriously. The result is supposed to be similar to that of the big bang theory, with the strings of our universe being created from this collision. They currently believe that the result is like the big bang in that initial conditions determine the outcome. However, this process is not currently understood and string theorists are struggling to explain this event in detail.

I propose that when this collision occured, the vast majority of energy from the momentum of the collision (over 90 percent as only 4 percent of the universe is comprised of ordinary matter) would be transferred to the brane itself in the form of vibrations. These huge membrane vibrations would be what we see today as dark matter (which is the curvature of spacetime). Approximately 4 percent of the collision's remaining energy would have been transferred directly to the virtual strings that were being generated within the membrane at the precise instant of the collision. This is because only about 4 percent of spacetime within the brane was in the process of generating virtual strings at the exact moment of the splash, creating ordinary matter and antimatter.[6] Our universe would be an average result of such a collision. This direct transfer of energy from the big splash's membrane vibrations to the virtual strings being created within our brane during the collision produced all the matter within the new universe.

This realization that energy was transferred directly from the brane collision to the individual virtual strings within our brane at the instant of the collision means that matter's mass is not caused by the closed-loop strings of current M-theory (which have no interaction with our brane). Instead, I propose that gravity is *the effect that a string's vibrations have upon the membrane within which it travels*. Imagine a string within a four-dimensional membrane as in fig. 8.2. If a string has the type of vibrations to act as matter, the string's vibrations in the three gravity dimensions will stretch the membrane around it as it travels, curving spacetime (as is consistent with relativity). Therefore, our membrane now *is* spacetime! In the original version of string theory before M-theory, according to Dr. Brian Greene, the higher the frequency of a string the more massive the particle it emulates.[7] This would also be true for the *membrane theory of gravity*. The more energetic the vibrations of a string of matter, the more those vibrations will stretch our membrane which the string is attached to, creating the force of gravity.

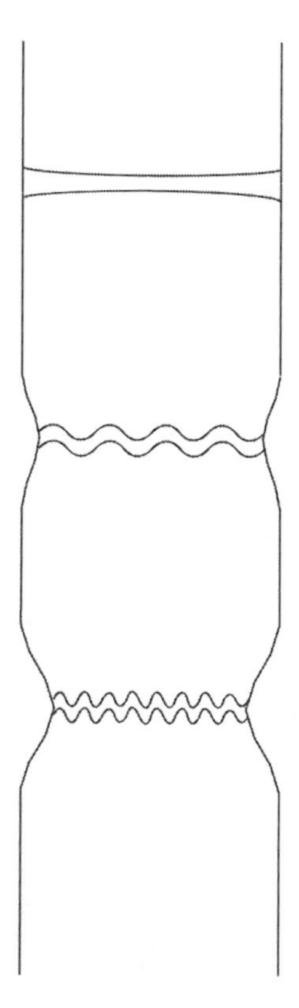

Fig. 8.2. *A representation of an "MRI view" of a membrane, strings, and their curvature of spacetime. In this sliced-open diagram, the lines on the left and right represent our membrane and its curvature. The three horizontal strings (hollow two-dimensional tubes) are (top) a virtual string that hardly vibrates at all and is basically a hollow tube that was separated from our brane by quantum noise; (middle) a string that represents a quark or other massive particle given energy from the big splash; (bottom) a string that represents a more massive strange, charmed, or similar quark given additional energy from a high-velocity impact with another string whose mass (vibrations) curves our brane even farther, creating gravity. The more periods the string has (that is, the more rapidly it vibrates), the more it contracts the membrane due to its vibrations and motion and the more massive the particle it emulates.*

With this modification of M-theory, it is no longer leaking loops of gravitons but string geometry and dynamics which create the effect of gravity. When a virtual string is disturbed into existence by quantum noise, it is a two-dimensional tube of our brane that is separated from our brane in its center but still connected to the brane on either end. While still virtual, the string has incredibly slight vibrations that have very little effect on our membrane. When struck by a big splash, however, a virtual string is given enough energy to vibrate rapidly and become a string of matter, such as an electron or a quark. As the string vibrates, the distance between either end of the string should decrease as more of the string's length is pushed farther away from the plane of the string's vibrations (see fig. 8.2). The stronger the vibrations of each string, the more that its ends will be pulled together, curving our brane and creating the "force" of gravity. Because of this, the more energetic a string's vibrations, the heavier the particle it emulates. Gravity is now a side effect of the vibrations of strings within our membrane. As with the previous method, when fusion occurs within a star, four hydrogen atoms are fused into a single helium atom, and some of the hydrogen's mass is lost in the process, having been converted into pure energy. When this occurs, one of the rapidly vibrating strings loses some energy by emitting a photon, and that string will vibrate less frequently, like switching from the bottom string diagram to the middle string diagram in fig. 8.2. This pushes the brane apart slightly, sending out a brane vibration (called a graviton) that has only the gravitational force and no other forces associated with it.

This explains gravity in a completely different manner than both string and M-theories currently do. M-theory states that gravity is a closed-loop string that can escape our membrane, accounting for gravity's weakness. However, the result of this is that the only way to satisfactorily explain how gravity works is by adding multiple universes and membranes from which gravity can leak to our membrane. Lisa Randall has postulated such a mechanism on the BBC TV series *Horizon*.[8] The chances that two randomly generated membranes would have this kind of symbiotic relationship are highly unlikely, because branes are only created so often and two of them would need to be created at exactly the same time and place, one within the other. The much more likely scenario is that membranes are independent and self-contained structures that were randomly generated from the quantum noise that exists everywhere in spacetime. Not a complex and highly unlikely arrangement of interconnected parallel universes as M-theory is beginning to require with its vanishing version of gravity. By eliminating the closed-loop strings of M-theory and replacing them with the simple and straightforward curvature of our membrane caused by the vibrations of strings of matter, we eliminate the necessity of relying on a contortion of parallel universes to explain the laws of nature. The universe is an elegant creation, even if its mathematics are turning out to be more complex than we may like. Simplicity requires we recognize that the most likely solution is *membrane gravity*, not an unlikely relationship between randomly generated branes.

The small percentage of big splash energy that got through to the virtual strings within each brane vibrated the three gravity dimensions of each string. This *directly* transferred energy from the splash to the nonenergetic virtual strings within each membrane, creating matter and antimatter quarks and electrons in up and down pairs as will be described shortly. The major brane vibrations that act as dark matter vibrate over large areas thousands of light-years across and will even form dark matter galaxies where these long vibrating brane strands cross each other. Huge chains of galaxy clusters will be created, attracted to these long strands of dark matter, and form an enormous lattice of strands following the brane vibrations and appearing similar to a three dimensional spider's web. This *membrane theory of gravity* would help to explain these huge structures as well. Because our brane was compressed during the collision due to its rapid vibrations, a network of crossing membrane waves would form appearing similar to a crystal lattice. These vibrations of the membrane itself would cause it to contract quickly into a much smaller volume as the membrane vibrates rapidly back and forth. It would form huge waves, both on the outer surface and throughout the inner structure of a four-dimensional membrane (see fig. 8.3). The dark matter resulting from the brane vibrations of the big splash could also cause the gases of the early universe to collapse and form stars and galaxies earlier than one billion years after the splash as we have observed, much more rapidly than current theory can explain.

Fig. 8.3. *A higher-dimensional view of two four-branes (with three spatial dimensions and one time dimension) colliding. As they collide, they create "big splashes" on either membrane radiating from the point of the collision. Both branes contract the most at the point of impact due to the violent vibrations. The vibrations would similarly reverberate throughout the center of our spherical four-dimensional brane, creating dense vibrating ripples of membrane (dark matter) throughout the new universe. The vibrations form a network of dark matter which helps to accelerate the formation of the new universe as the young gases are attracted to these dense, vibrating, dark matter ripples. Additionally, this collision starts off our universe at a size much larger than previously theorized, doing away with the need for any singularities. Therefore, physics never breaks down in this model of our universe, unlike the big bang model where all of physics breaks down before* 10^{-43} *seconds into the expansion from the so-called singularity. The universe simply contracts from the impact, and then starts to expand from a point possibly billions of light-years across. Even without the larger starting volume, our current minimum diameter is at least 156 billion light-years across, according to the WMAP science team.[9] These branes would not appear smooth before the collision as they do in this illustration (for simplification). Instead, they would ripple slightly due to quantum noise, and those ripples would be imprinted on our universe. They appear as areas of higher-and lower-energy impacts where the ripples constructively and destructively interfere with each other at the instant of the big splash. This pattern is visible in the cosmic background radiation left over from the splash as we will see in fig. 8.5.*

As our universe started to expand after the splash, the original gases cooled down and started to condense due to gravity. They were attracted to the network of dense dark matter gravity created by the massive brane vibrations left over from the splash due to *membrane gravity*. This accelerated the gases' collapse into immense, dense clouds. This network of dark matter helps to explain how galaxies formed so soon after the creation of the universe, with fully formed galaxies created in under one billion years which the standard big bang model has difficulty explaining. As the first huge gas clouds began to condense, gravity became so intense that atoms of hydrogen at the center of the clouds were fused together into helium and eventually into much heavier elements. This fusion releases a huge amount of energy and creates the first massive stars and eventually even black holes in our universe as these first stars die and collapse in on themselves. These first black holes swallow huge volumes of gas so rapidly that as the gas falls in toward the black hole, it is accelerated close to the speed of light just before getting sucked into the event horizon of the huge black hole. The friction of the gases circling the black hole at such high velocities heats the gases up to millions of degrees, which is hot enough to release X-rays and create intense interstellar winds. These winds push much of the galaxy-sized clouds of gas far away from the black holes at their centers, causing the clouds around the hole to collapse into a multitude of individual stars.[10] These stars very rapidly begin to fuse hydrogen and create the first new galaxies in our universe.

The first stars created in these new galaxies, in turn, create heavier elements due to the fusion of hydrogen and helium gases, and then the fusion of heavier elements such as carbon, nitrogen, and oxygen. If the star is massive enough, when it tries to fuse elements as heavy as iron, gravity is no longer strong enough to fuse the iron atoms together. Instead, the star collapses in on itself when fusion stops, and the star dies in a spectacular supernova explosion; fusing all elements heavier than iron together. This includes every heavy element all the way up to uranium. This explosion also disperses these heavy elements across the galaxy. These early supernovae send more shock waves throughout the gases of the young galaxies, causing the gases to condense in even more areas, creating yet more stars. Such processes continue for nearly fourteen billion years until we have the universe that we see today with nice, comfortable solar systems containing the heavy elements within them that life on Earth (and elsewhere) needs in order to grow, survive, and thrive.

The vibrations of a membrane are not contained within a small, confined space as regular string's vibrations are (the strings vibrating within our universe that act as matter and energy and are confined to the Planck length). Because brane vibrations have our entire membrane to expand throughout (which could be much larger than just our visible universe depending on its shape), their vibrations are not stable and will dissipate throughout our entire brane. They do so according to the inverse square law as the waves expand in three dimensions and dissipate throughout our universe and beyond into the rest of our brane (accounting for gravity's weakness). This is another advantage of a *membrane theory of gravity.* It explains gravity's weakness compared to the other forces of nature as a fundamental property of spacetime without the need to resort to vanishing strings or other unlikely processes. As our universe expands, the dark matter brane vibrations will slowly be converted into dark energy as our brane vibrations relax and dissipate, pushing points that were close together farther apart (see fig. 8.4). This will happen as the vibrations of our brane occur less frequently, especially in the empty (starless) regions of the universe between galaxy cluster strands. The cooling of our universe will start to convert the potential energy of our vibrating membrane into kinetic energy. This release of stored vibrating brane energy pushes our universe apart, causing it to steadily accelerate, beginning slowly then increasing in speed until it reaches its original size before the splash, at which point the acceleration will decrease and eventually end (as in fig. 8.4). In this manner, *membrane gravity* explains both dark matter and dark energy simultaneously, where current physics and string theory cannot explain either one. This hypothesis is supported by recent studies measuring aspects of dark matter. It was found that dark matter behaves like brane waves as the average nearby dark matter "wave" averages one thousand light years across. These waves have an average density of four hydrogen atoms per cubic centimeter, giving them an even density, just like brane waves would.[11] These measurements can be used to validate membrane gravity because if we look further back in time and find that dark matter waves are smaller and have a higher energy density, it will help to confirm *membrane gravity.*

In the areas of space where galaxies are plentiful, stars are constantly converting matter into energy as they fuse elements together in their cores. This act causes our membrane to vibrate slightly as it creates gravitons from the matter being fused together, and some of that matter is turned into massless photon energy. As a string of matter releases its tension on the third gravity dimension when it is converted into light, the dimple once created in the membrane by that matter is released. When it is, the brane vibrates back and forth, creating tiny gravity waves which we observe as gravitons. We see them as gravitons because these vibrations of our brane are only Planck-length in size, and we have no way to discern something that small as anything other than a particle at this point because of our limited vision.

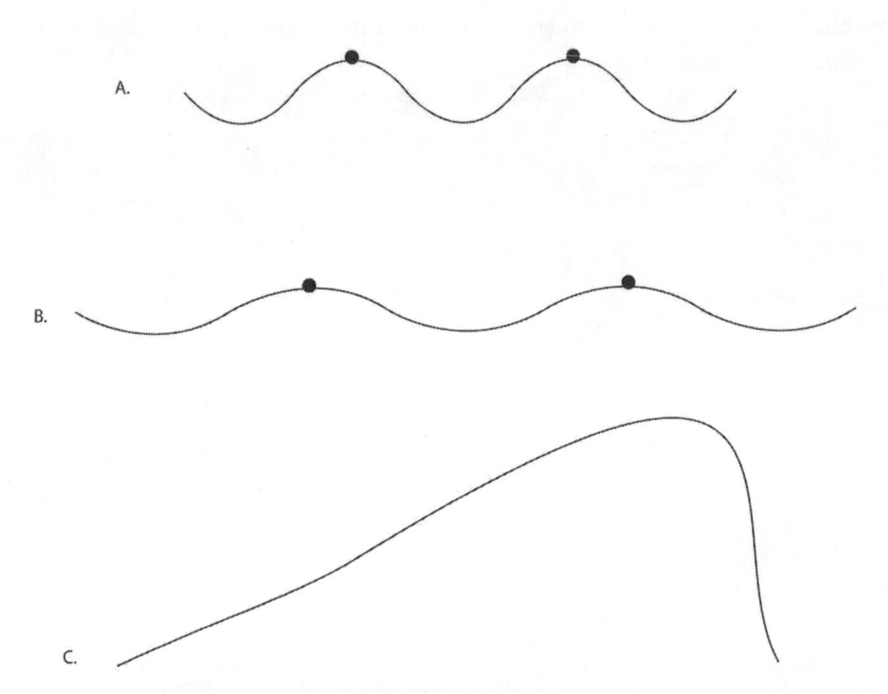

Fig. 8.4. *Vibrations relaxing and the acceleration of our universe's expansion due to membrane gravity (not to scale). As our membrane relaxes, two points that were closer together (A) get pushed farther apart (B) as the spacetime between them vibrates less frequently and becomes less compressed. As this occurs throughout our universe, it expands more rapidly until the brane relaxes almost completely, at which point the expansion slows down. Because of the rate at which brane vibrations relax, spreading out spacetime, dark energy (C) the rate of acceleration of the universe should appear relatively constant in the beginning. Our universe's acceleration will begin slowly after the big splash due to the brane's rapid vibrations, shrinking the size of our brane itself. Our universe will continue to accelerate, appearing as constant acceleration, until our brane expands completely, and all brane vibrations have ceased. No other theory predicts that dark matter is converted into dark energy, or that our acceleration will one day end as out universe reaches its original size as this membrane theory of gravity does.*

These constant vibrations from gravity waves and gravitons could help to keep the dark matter of our early universe in its vibrating state and acting like matter around galaxy clusters. As long as these brane vibrations continue, dark matter could maintain its hold on our universe's galaxies and galaxy clusters. In the empty voids of space between galaxy clusters, the brane waves are far from the spider's web of massive brane vibrations created in the splash, and relax the most rapidly as temperatures in those voids is falling close to absolute zero. In these regions of space, the membrane vibrations left over from the big splash are being converted into dark energy the fastest, pushing spacetime apart as they do.

Quasars are galaxies created very early in our universe which have young, active black holes at their centers that feed on enormous amounts of hydrogen and helium gas. Some quasars may even have several black holes orbiting each other which could eventually merge into a single huge hole. As the gases around these holes spiral in toward their event horizons, the gases accelerate close to the speed of light, heating up to millions of degrees before the black hole swallows up the gas. At the last instant, the holes send massive gamma rays and beams of highly accelerated gas shooting out from both poles of the black hole. The massive gravitational energy may have aided in sustaining our membrane's rapid vibrations during the early universe, especially near galaxies, holding them together. Examples of such energy include gravity waves being produced from orbiting black holes, supernovae, and hypernovae, among other graviton and gravity wave sources. Hypernovae are created when very large stars collapse instantaneously into a black hole when they die due to their enormous mass.

As our universe ages, the expanding empty (starless) spheres of space between galaxy clusters' strands begin to have very few gravity waves passing through them at the same time our brane relaxes even further. This brane relaxation begins to accelerate the expansion of our universe. These regions are the most isolated from the network of brane vibrations created during the big splash and from any gravity wave sources within faraway galaxies. Because of this, any vibrating membrane waves in these regions begin to relax and dissipate by this time. The relaxation of original brane vibrations in these regions causes their dark matter vibrations to be converted into dark energy expansion forces. The brane relaxes and spreads back out the most in these regions first, as they will elsewhere throughout our brane as its vibrations continue to relax and accelerate our expansion.

This *membrane theory of gravity* could also explain other mysteries of our universe: why gravity is so weak and why so little of our universe's mass is made up of normal matter and energy that we can see. Our membrane began life as a single four dimensional string (one dimension being time). This brane eventually inflated to a size much larger than just our own universe. When our brane was young and small, its interactions (if any) with strings of matter could have been much more powerful, similar to the electromagnetic interactions of strings today. As illustrated in chapter 1, jumping up and down can effectively illustrate that the electromagnetic force is 10^{39} times more powerful than gravity as your feet are stopped instantaneously by the ground when the electrons in your feet and the ground meet.[12]

As our brane grew from the size of a string to a size larger than a galaxy and then larger than many galaxies, its surface area inside and out continued to expand. According to a *membrane theory of gravity,* as it did, the effect of vibrations of standard Planck-length strings of matter (which curve our brane) got continuously weaker and weaker as those tiny strings stayed the same size while our brane continued to expand. The gravitational effect that only one of our strings of matter such as a quark or electron has on our brane continued to decrease as it had a smaller and smaller percentage of the membrane's surface area to vibrate against. When our brane finally reached its current size, the vibrations of strings of matter have an infinitesimal effect on the tiny percentage of the brane that they vibrate against. Therefore, *gravity's weakness should be directly proportional to the size of our membrane.*

Because of this, it may be possible to calculate the approximate size of our membrane by comparing the strength of gravity to the electromagnetic force (as they may have been comparable when our brane and a single string of matter were of equal size) and then by factoring in the distance that the electromagnetic force acts over. The approximate distance between a proton and an electron in a hydrogen atom (the Bohr radius) is about $5 * 10^{-11}$m. If we multiply this by the difference in power between the two forces (10^{39} times), we get $5 * 10^{30}$ m. A light-year is $9.5 * 10^{15}$ m, so if we divide our result by the distance in a light-year, we get $5.26 * 10^{14}$ light-years, or a maximum radius of 526 trillion light-years across (roughly). If our universe is at least 156 billion light-years across as a recent WMAP science team study found[13] (and possibly larger if it began in the manner described earlier), then our membrane could be up to six thousand times larger than our universe currently is.

A *membrane theory of gravity* could also explain why normal matter is such a small percentage of the mass of the universe. This would be because only about 4 percent of spacetime within our brane was in the process of generating virtual strings at the instant of the brane-brane collision. These virtual strings created an outlet other than the brane itself where the energy of the big splash collision could go, vibrating the virtual strings and creating matter. Any energy not transferred directly to the virtual strings within our brane would have simply vibrated our membrane itself, creating the vast majority of mass in the universe as dark matter. If there were some strings left over from previous collisions of our brane in addition, they would also be available to be vibrated into matter and antimatter strings during the collision.

This method of universe creation would also explain why only matter and not light was created directly in the big splash. Where virtual strings existed within our brane during the moment of impact of the big splash, all those strings were struck in their three gravity dimensions. This occurs similar to how a piano mallet strikes a note, causing the piano's string to vibrate at a particular frequency. Because this impact influenced all three of our string's gravity dimensions at once (as those are the dimensions that interact with our brane), it caused the strings to vibrate in at least all three brane curving dimensions of gravity. This created strings with mass such as quarks and electrons in matter and antimatter forms, depending on the strength of the impact where each pair of virtual strings existed. The collision could not create photons directly as all three matter dimensions were impacted, and light only vibrates in two of our three gravity dimensions. However, light would soon be created as matter and antimatter strings annihilated each other, creating photons. Possibly milliseconds later, this "string soup" would condense into matter, emitting photons of light which become the cosmic microwave background radiation (CBR) . From this point forward, this "big splash" behaves exactly like the big bang model.

A *membrane theory of gravity* could also change our idea of what multiple universes are within other membranes. Current theory states that the initial conditions of a big bang or big splash determine all the physics of the universe created from it. It is currently believed that other universes will all be very different, and that it just happens to be luck that initial conditions were so favorable for life in our universe. However, if this *membrane theory of gravity* is correct, all universes created within four-branes (with three spatial dimensions and one time dimension) like ours will have nearly identical laws of physics. Only the *strength* of gravity will differ, depending on the size of the brane the universe is created within. The smaller the brane, the stronger the force of gravity; and the more powerful the collision that creates the universe, the more energetic the matter and antimatter strings that will be created within it. If a brane collides more gently than ours did, less matter and hardly any antimatter would be created within it. If the brane were smaller than ours as well, its stronger force of gravity could compensate for the smaller collision to create a viable universe, habitable by creatures like us. If, on the other hand, an enormous four-brane has a minor collision with another brane, the small amount of matter created (and its weak gravity) would make for a universe of diffuse gas that may never form a single star. In five-brane, six-brane, and higher spatial dimensional universes, however, the laws of physics would be completely different but the same for each number of dimensions (except for gravity which would again vary on brane size and the number of spatial dimensions each brane has).

If this *membrane theory of gravity* is correct, it should be *verifiable through observation* using several methods. We could calculate the average mass of space in the universe during the first eight billion years of its existence (mostly in the empty-of-stars regions of diffuse gas between galaxy clusters) and compare those results with the average mass of empty space in the universe during the last five billion years. *If our results show that dark matter has been disappearing as our universe's expansion has been accelerating (and losing mass during the last five billion years,) it would confirm this membrane theory of gravity and validate string theory simultaneously.* Gravity would no longer be leaking from our universe as M-theory first predicted or originate from a membrane or dimension just slightly away from our own (as a newer string theory of gravity postulates). Another way to confirm the *membrane theory of gravity* is by measuring dark energy. If we find that our acceleration will eventually end when our brane expands completely, and will not simply expand forever, this will confirm brane gravity. As our membrane's vibrations left over from the big splash relax, any two points on those waves will be pushed apart slowly at first and then more rapidly. Finally, the acceleration will slow down again as our brane expands out toward its original size before the collision.

This theory also shines new light on the cosmic background radiation (CBR). Because the collision of branes struck our universe's strings in our gravity dimensions, it transferred energy directly to the virtual strings creating a vast majority of normal matter in our universe compared to antimatter. In those areas of higher energy collisions where brane ripples overlapped, generating constructive interference (and more energy), some higher energy antimatter, such as positrons (positive electrons), was created in those areas. This hot bubble of subatomic strings created by the collision expanded, and the antimatter created in the collision was instantly annihilated when it interacted with its normal matter counterpart quarks and electrons, generating gamma rays. As the bubble of subatomic strings went through a phase change as it cooled and condensed from hot quarks into cooler matter such as hydrogen and helium, it emitted a flash of photons which we see today, 13.7 billion years later, as the weak microwave cosmic background from the instant our universe condenses into normal matter from the string soup (or quark soup, if you prefer), the "big splash" behaves exactly like the "big bang" theory does.

The Wilkinson Microwave Anisotropy Probe, Image courtesy of NASA, thanks to the WMAP Science Team.[14]

Fig. 8.5. *NASA's WMAP satellite detected the millionth of a degree difference between the hotter, lighter areas of the cosmic background radiation and the cooler, darker areas. What was once a flash of photons shortly after the birth of our universe—when it condensed into hydrogen and helium gas—has cooled and been stretched over the last 13.7 billion years as our universe has expanded. This has turned light rays into the weak microwave background radiation seen in every direction of the sky, which is the fingerprint of the creation of the universe.*

Because the universe could have been created at such a large size with the *membrane theory of gravity* (possibly billions of light-years across), the quarks and electrons created in the collision may have condensed almost instantly (or at least much more rapidly). They condensed into hydrogen, helium, and deuterium atoms (similar to how water goes through phase changes as it cools from gas to liquid to solid). This would mean that the cosmic background radiation may not have been created four hundred thousand years after the big splash as is currently thought.[15] Instead, the CBR could have been created very shortly after the splash and much more rapidly than the "standard model" predicts. Our universe would have rapidly cooled and coalesced from hot quarks and electrons into hydrogen and helium atoms due to the comparatively large size of the early universe.

This would mean that when we look at the CBR, we could be looking *almost directly at an image of the creation of the universe itself.* We would no longer be looking at a picture of our universe nearly four hundred thousand years later as current cosmologists believe (according to the "standard model" of the big bang). With this new *membrane theory of gravity*, the universe could have started at a much larger size, possibly several billion light-years across, and started expanding from that point. According to the WMAP science team, after 13.7 billion years of expansion our universe is at least 156 billion light-years across and possibly even larger, depending on the starting size of our universe.

Another recent piece of evidence in favor of this membrane model of gravity is the Hubble Space Telescope study of 2003 by quantum gravity physicists. They found that the universe is not as "lumpy" as it should be if spacetime were pixilated at the Planck length of 10^{-33}cm (the smallest anything can be according to string theory, as well as the length of a single string). Quantum gravity physicists believe that photons traveling through our universe would move from "pixel to pixel" of Planck-length segments.[16] Images of objects that are farther away should have appeared smeared compared to nearby objects as more distant photons would have to travel through more segments. However, these results were not found. Photos of both nearby and distant objects appeared equally clear. Instead of casting doubt on the Planck length or other solid physics, it is further support for the *membrane theory of gravity* that photons are not traveling through "pixilated" space. Rather, they are traveling through a perfectly smooth membrane, which would not blur photons in the least.

We need to move from a particle-based view of the universe to a membrane and string-based view. Researchers are having trouble figuring out dark matter, dark energy, and many other recent discoveries because they view the universe as a collection of particles within a vacuum with "nothing" beyond the edge of our universe. Even M-theory currently suffers from this view as it still does not see our membrane as the source of gravity, with each string currently behaving like a particle in standard physics. The closed-loop graviton string is what is holding M-theory back from becoming a complete theory of physics. The picture of the universe becomes so much clearer when we realize that we live within a fluid-like membrane that may extend beyond our visible universe. Because of this, all matter and energy are simply vibrating strings, formed from and connected to our membrane, that travel through our brane and whose vibrations generate all the forces of nature that we observe.

Additionally, inflationary models of the big bang have been under tighter scrutiny in recent years as results from WMAP have been analyzed, and the results do not quite match predictions. This is a very difficult feat, considering there are so many different predictions. There are many problems with inflationary models, one of which is quite similar to complaints about the first five string theories before they were unified under M-theory. There are so many different inflationary approaches, with such a wide range of predictions, that it has been suggested inflation could never be disproved by observation![17] Here is another problem with inflationary models: What caused the inflation and the "bang" in the first place? This question can never be answered with current models, which is very frustrating to say the least. Some physicists seem not to mind being unknowledgeable of what event actually triggered our universe's creation and expansion. However, it seems vital for any comprehensive theory of our universe to be able to predict how our universe began and how it will end.

In a *Science* magazine article by Adrian Cho, the latest finding by Christoph Adami, of the California Institute of Technology in Pasadena and the Keck Graduate Institute in Claremont, that quantum entanglement is linked to gravity could be one of the strongest pieces of evidence yet for the *membrane theory of gravity*.[18] If two particles such as electrons are entangled, an observer could manipulate one of the particles and cause it to spin up. (Electrons have two types of spin: one called "up" and the other, "down.") When the particle spins up, they would instantly know that the other entangled particle's spin is down, no matter how far away it is. This interaction makes sense in a *membrane theory of gravity*, because all matter and energy in our universe are connected to each other through our brane. The strings are physically connected to the brane, and their gravitational vibrations curve our brane. Quantum information could only be transmitted or linked if the particles themselves were connected or linked in some way, as they would be through our brane. All matter in the universe—and even light itself, which vibrates in only two gravity dimensions and one higher dimension—is moving through our brane, constantly in contact with it. No other theory can yet account for why quantum entanglement would be linked to gravity, especially current M-theory where gravity is not linked to anything and passes right through our brane as if it were not even there.

Although classical physics can explain light's effects, a membrane model of gravity could make them easier to comprehend. For example, why does light, a massless "particle," exert a tiny force on matter? Solar sails take advantage of this property of light when trillions of photons bounce off a reflective lightweight Mylar sheet and cause acceleration. Because light has no mass, this would seem an impossible task. However, by using M-theory with membrane gravity, we can see that because strings of light vibrate in two of our three gravity dimensions, curving our brane in those two dimensions, light waves can bounce off the brane dimples of the Mylar atoms in those two dimensions. This action transfers a very slight amount of the light string's frequency vibrations to the Mylar as momentum and decreases the light's vibrations (frequency) very slightly from the energy transfer (see fig. 8.6). This would also explain how light can transfer its energy as heat to other substances. Light causes the molecules to vibrate faster after a collision, because the faster an atom vibrates, the higher its temperature.

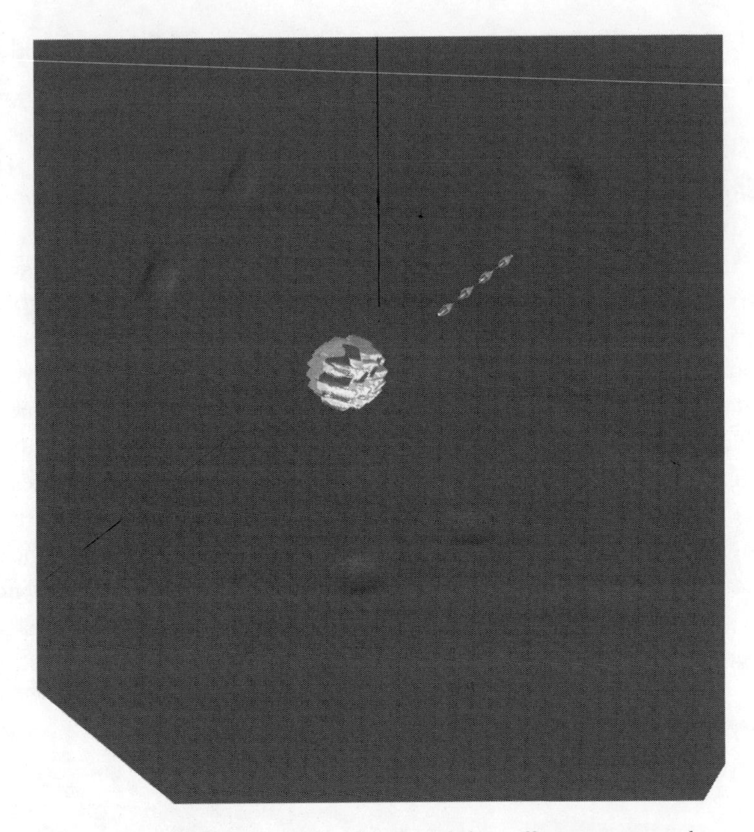

Fig. 8.6. *Strings of an electron and a photon and their effects on our membrane and each other (exaggerated, not to scale, dimples inverted). The round string on the left represents an electron or quark, which is matter that vibrates in all three of our gravity dimensions. This curves our membrane, creating dimples in all three dimensions (the three dimples pictured below it, to the left, and to the right). The photon of light string (the line with four bulges above and to the right of the string of matter) vibrates in only two of our three gravity dimensions, plus a higher unseen dimension. Therefore, light curves our membrane in only two of its three dimensions (the dimples below it and to the left of it only—the dimple on the right wall belongs to the string of matter, not the photon) and thus has no mass. However, because a photon does curve our brane in two dimensions, it can interact with matter in those two dimensions and transfer vibrations from the photon to the electron through colliding with the electron's brane dimples. Because gravity is so weak, and light only interacts with matter in two of the three gravity dimensions, the energy transfer is extremely weak, taking trillions of photons to have any noticeable effect on a lightweight reflective sheet of Mylar.*

A *membrane theory of gravity* could also explain why $E=MC^2$. As a string of matter vibrates, it curves our membrane in three dimensions. However, when a string of matter is converted into energy, the string will only vibrate in two of the three gravity dimensions, transferring the energy from the one gravity dimension removed into the two remaining gravity dimensions, conserving the string's energy. Therefore, the string's energy remains constant as it vibrates in fewer gravity dimensions that we can see. $E=MC^2$ simply illustrates the conservation of energy as the vibrations of a string in three gravity dimensions are compacted into the vibrations of a string in only two gravity dimensions that travel at the speed of light as matter is converted into energy.

This theory could also explain why all fundamental strings have two types of spin—up and down—as discussed earlier. If, when the big splash occurred, the only strings available within our brane to which energy could be transferred were virtual strings that existed at the precise instant of the collision, then the energy of the collision would have been transferred directly to each couple of virtual strings. Each pair of strings would have been given the same amount of energy, depending on their location within the brane. One pair, if given enough energy to become electrons, would split apart as their electromagnetic vibrations repelled each other, and one of the electrons would spin up, and the other would spin down. This occurs because the waves of each virtual string in the pair vibrate in the opposite direction of the other, because the pair would cancel each other out when they joined back together and disappeared had they not been struck by the big splash.

The same reaction would occur with virtual strings given enough energy to become quarks. The pair of strings would be vibrated, forcing each pair apart into up and down quarks (with opposite charges of +2/3 and-1/3, respectively). The quarks would also be "entangled" as they were born from the same pair of virtual strings. After being created, the quarks would join together with two other pairs of quarks to form a proton and neutron (with two up quarks and one down quark and two down quarks and one up quark, respectively, in each). This could also account for why some deuterium and tritium was created in the big splash. Many protons would have attracted neutrons to them (being created nearby) before an electron was captured by the proton creating the first atoms in our new universe.

This method of creation for strings could also explain why electrons (when cooled sufficiently to reduce their electromagnetic repulsion) pair up into up and down sets and flow without resistance through superconductors at very low temperatures. Because up and down electrons were created from opposite wave virtual strings, if they are brought close enough together by cooling them, the electrons will fall toward one another and travel together. This is similar to how gravity's curvature of spacetime causes objects to fall toward one another. Because the electrons pair up, they no longer bounce randomly through the material they travel through as they normally would, heating it up and causing resistance. Instead, they form orderly pairs and flow without resistance and with zero loss of energy, a property we can take advantage of to transfer electricity without loss and to create powerful electromagnets.

It is impossible to tell how many membranes have yet evolved like the one our universe was created within, but the birth of a new universe could be a daily or even hourly event depending on how many branes have evolved so far. If the *membrane theory of gravity* is correct, we likely live within a multiverse (a set of multiple possible universes) of branes floating around in higher dimensions. This system of strings and branes continues to evolve, getting ever more complex and building upon itself. Complex chemical reactions made possible from the heavy elements created within the first stars (and dispersed throughout young galaxies when these stars die and explode) will continue increasing the complexity of this system. This will lead to the creation of complex organic molecules. After billions of years of replication, competition, and evolution, these once simple molecules will form beings of such intelligence that they can look back on the universe and actually understand where it came from, how it evolved, and how it will end. The universe really is a thing of beauty.

According to the *membrane theory of gravity*, our universe will end in ice rather than in fire as its expansion is currently accelerating. However, our universe will not continue to accelerate its expansion *forever* as some astronomers currently believe. As the universe continues to expand and our membrane's vibrations continue to relax, dark matter is being constantly converted into an expansionary energy (dark energy). As our brane continues to expand (back toward its original size before the splash), the more that dark matter (brane vibrations) will evaporate (relax). Our universe will end up being a cold, dark place, but the galaxies still remaining in our ancient universe will slow down their expansion until our brane expands back out to its original size. The last of the stars will eventually die out, and long afterward the remaining black holes will evaporate. Then this universe will be ready for the next membrane collision to start the process all over again.

We could travel anywhere within our brane, using SlipString Drive, and possibly someday even peer outside of it using similar technologies. It should also be clear by now that our own universe is finite, not infinite, being contained within a single membrane of which there are a multitude. The multiverse, however, may or may not be infinite. It is big, but do not let that fool you. If there are still strings and branes being created out there today, then the multiverse could be finite as well, just very large and still growing.

These modifications of M-theory avoid the apparent breakdown of physics we encounter during the creation of our universe at times before 10^{-43} seconds after the big bang, when the entire universe was supposedly compressed into a singularity. This new version of the big splash with a *membrane theory of gravity* does away with all the breakdowns of physics, infinite energies, and infinitely small sizes of the old big bang model. This leaves a much more stable and robust theory for the creation of our universe. It also predicts dark matter and dark energy and how they change over time which neither the standard model or M-theory can currently do. It also explains why gravity curves spacetime and why spacetime acts similar to a fluid (as branes do). It predicts a "flat" universe (as we have observed) and among other recent observations it explains, it illustrates how quantum entanglement is linked to gravity, and it behaves just like the big bang after our universe goes through its final (and only) phase change into matter. Once rigorously tested, this hypothesis could finally give physics a true Theory of Everything and successfully combine relativity with quantum mechanics through M-theory and this *membrane theory of gravity*.

CHAPTER 9

Life, the Universe and Everything, or: How the Universe Will End...But Must We End With It?

When our universe reaches about a hundred billion years old (or somewhere around there), the last generations of stars will have used up all their fuel and will begin to die out. As they do, the universe will become dimmer and dimmer, until finally, the last star burns itself out. Nearly all the matter in the universe will end up falling into black holes by this time. All local groups of galaxies will have merged into single utterly massive galaxies separated by hundreds of millions or more light-years. With the mass of at least ten or twenty Milky Way galaxies, these behemoths will funnel almost all of their mass into the black holes at their centers or the black holes orbiting where their stars used to be. The only "normal" matter remaining will be a few burnt-out stars and their cold (near absolute zero) planets still orbiting them. As more time passes, the last of the universe's dark matter membrane vibrations will become so dispersed and weak that they barely make our brane flutter. This loss of dark matter will cause the last remaining dead galaxies to slowly disassemble, dispersing their black holes and cold dark physical matter thinly throughout our membrane.

According to Stephen Hawking, black holes will evaporate as they age (after they finish gobbling up any matter falling into them). This evaporation takes place as virtual strings pop in and out of existence near the event horizons of black holes (as they do everywhere else in the universe). These "virtual particles" (strings) act as matter, but one string is the inverse vibration of the other, which cancels out both strings when they join back together and disappear from existence. If these virtual strings come into existence near the event horizon of a black hole, and one virtual string of the pair falls into the black hole, it will collide with the tiny ball of strings at the core of the black hole. This will annihilate one of the black hole's strings of matter. Because of quantum entanglement, the other virtual string will be given the quantum state of the annihilated string from inside the hole and be accelerated away from the hole's event horizon, escaping its gravity. When this happens, there is a net loss of matter from the black hole as a faint glow of "Hawking radiation" is emitted from the dormant black hole.[1] This evaporation over time means that all matter within the black hole will eventually be recycled back into the universe as weak Hawking radiation. A googol (10^{100}) years from now (although probably sooner as our membrane should continue to expand over this time, reducing the strength of gravity), the last of the black holes will evaporate. They should actually evaporate much more quickly than previously calculated due to the fact that their cores of compactified strings are tens of meters across. The size of the core would accelerate evaporation compared to that of a black hole with a singularity at their core. As these black holes dissolve to the mass of about thirty suns or less, they will no longer have the mass to overpower the vibrations of the individual strings within them. At that point, the hole will release all its remaining matter in an enormous explosion. By that time, our universe will just be one huge, cold, smooth, and fairly boring sheet of weak radiation.

After some time, and possibly before our universe gets too old (possibly around 100 billion years), a new universe could be created within our brane. This would occur when the brane our universe is within collides with a different brane, starting the process all over again. Our brane is floating away from the point of its last collision (which created our universe), so it is only a matter of time before it strikes another, starting the process all over again! Originally, I thought that we would need to create bubbles of strings around space ships in order to leave our universe for a younger, healthier one when ours grew old and dark as other string theorists also postulated. After my revision of M-theory with the *membrane theory of gravity* to explain gravity, dark matter, dark energy, and other phenomena, I realized that our membrane *is* spacetime (the medium that we travel through). Because of this, there is no need to create a bubble of strings around us in order to leave our universe. Our brane *is* one huge string! We will be able to create gravity wave ripples and travel anywhere within our brane in this universe and beyond at incredible speeds using SlipString Drive.

In order to avoid being recycled into our new universe, it may be possible to figure out a way to survive a "big splash" event. We could scatter a multitude of SlipString probes across more than nine hundred trillion light-years of our membrane to report if or when a new universe was created or to report back regularly. If several probes in a particular region did not report back, then we would know a collision had taken place in the missing probe's region, and it was expanding toward us at the speed of light. If a big splash occurred where a probe was, all the probe's strings would be vibrated to higher energy levels (such as antimatter), destroying the probe. We could also design probes with periscopes, as in chapter 3, which would survive (even if the tip of the probe's periscope did not) so the probes could report their status back home. With this information, we could plan an expedition to the new universe as soon as it would be safe to do so when our universe was at or near its end. Since we most likely live within a spherical brane, we may need to keep a percentage of our ships completely isolated from spacetime just in case our brane is struck without warning, so that at least some of our species will survive another big splash. However, even in this scenario we could be warned by periscope probes of an impending big splash wave headed toward us, and prepare to travel through the wave to a point on the other side that has condensed and cooled into a habitable universe. Whatever the shape of our brane, with enough probes to warn us of a new splash on the way, humanity should be able to survive the creation of a new universe and live on to explore that universe from its very beginning.

When it is time for us to leave our universe, we could head for our new home at SlipString speeds in our gravity wave-isolated regions of spacetime. We would just need to stock up on plenty of fuel for our fusion or antimatter engines, which we could easily do before our journey. It would not be difficult to create moon-sized fuel ships full of enough gas to last for millions of years with the technology to isolate large volumes of spacetime, negating gravity. It should not be too many years of travel time on SlipString "arcs" before we reach a new universe and are able to settle down for another several hundred billion years there. If the *membrane theory of gravity* is correct, and gravity is not transferred by closed loops of strings that can escape our membrane as M-theory currently postulates, then it is doubtful that we will ever figure out a way to leave our current membrane if necessary. Because no strings can escape our brane, it is likely that we could not use our gravity wave drives or any other similar technologies to push ourselves off of our current membrane and onto a different parallel brane to explore a parallel universe. However, there is still a good chance that we could survive the next big splash and then explore a brand new universe which we could call home after our current universe has been recycled in the splash. Good luck to future generations on the expedition, and happy universe hunting!

Appendix Preface

This appendix is a self-contained article written for string theorists to use and to quote in their papers and articles concerning the *Membrane Theory of Gravity*. It is written in the hopes that it will be used as a guideline to develop the M-theory and D-Brane mathematics necessary to describe membrane gravity and to develop M-theory to its full potential as a final theory of physics.

Appendix
A Hypothesis for a Membrane Theory of Gravity

How to Unify All Forces, Explaining Cosmological Conundrums Including Dark Matter and Dark Energy by Modifying Gravity in M-Theory

by Andrew L Bender

A Hypothesis for a Membrane Theory of Gravity

Andrew L Bender (andrew@slipstring.com, http://www.slipstring.com)

Abstract: Proposed is a modification of gravity in M-theory that could successfully combine relativity and quantum mechanics by means of a *Membrane Theory of Gravity.* The force of gravity would be produced by the curvature of our membrane by the vibrations of individual strings of matter in the three gravity dimensions of M-theory. Gravity's weakness would be due to the size of our membrane relative to the Planck length of an individual string. If the forces of electromagnetism and gravity were roughly equivalent when our membrane was the Planck length, then as it expanded to its current size, the strength of gravity would have decreased in proportion to our membrane's size, roughly calculated to have a maximum radius of 526 trillion light-years. This theory allows for a direct method of energy transfer from membrane collision to individual "virtual" strings present within the brane at the instant of the collision. This explains the creation of our universe without a singularity while keeping physics intact throughout the big splash. Dark matter and dark energy are revealed to be vibrations of our membrane left over from the original collision that created our universe. The relaxation of those vibrations (whose potential energy is converted into kinetic energy) causes spacetime to expand more rapidly as it ages and to accelerate at what appears to be a constant rate until all dark matter relaxes completely, stopping the acceleration. Additionally, several methods of observational confirmation of this theory (and therefore M-theory as well) are proposed.

M-theory has been a significant development in string theory and in physics in its relatively short lifetime. M-theory encompasses five earlier competing string theories, all of which had postulated ten dimensions and had seemed perfectly plausible. However, five different theories were "too much of a good thing" as, ultimately, there can be only one correct theory of everything (TOE). In 1995, Edward Witten came up with an elegant solution to the multiple theories problem.[1] By adding an additional physical dimension even smaller than the other nine physical dimensions (the remaining dimension being time, thus accounting for the total of eleven dimensions), string theory finally seemed to fall into place. The five previous versions melted together and were akin to looking into a five-segmented mirror. Each image appears slightly different, but the subject looking in the mirror stays the same. Thus, M-theory was born, and string theory was saved (or so we thought).

The additional dimension turned strings (the most fundamental form of matter and energy in the universe) from one-dimensional vibrating loops of energy into two-dimensional vibrating tube structures with no thickness whatsoever, just surface area. The vast majorities of these strings were no longer loops but were open-ended with both ends of the strings being connected on either end to a membrane (or brane for short—a new result of the eleventh dimension). The branes were the result of these new two-dimensional strings that had the possibility of inflating and stretching into huge sheets or into a torus among many other multidimensional shapes now allowed by the eleventh dimension and the mathematics of brane dynamics. The strings that make up matter and energy are connected to our brane (which has three spatial dimensions and a fourth dimension for time), connecting our universe to the brane. In the original M-theory, however, gravity was a closed-loop string and not connected to our brane. Therefore, it could leave our membrane and leak from our universe, accounting for gravity's weakness relative to all other forces.

Physicists are currently searching for "disappearing gravitons" in atom-smashers in order to prove or disprove this new theory. Assuming M-theory is largely correct (and a brilliant evolution of string theory), I believe it is unlikely that any gravitons will be disappearing from our membrane, as experiments should eventually confirm this. Our universe is a highly elegant and efficient creation, and leaking gravitons just do not seem to fit squarely into that elegant universe. The first problem with these loops of leaking gravitons, and therefore current M-theory, is that it has no mechanism for energy transfer from a membrane collision directly to the virtual strings within our brane. Other problems include its failure to predict dark matter and energy and having to rely on parallel universes in order for gravity to work properly. (Currently, gravity must leak to our universe from a parallel external brane in order to work properly.) The odds of two branes having such a relationship by random processes are extremely low. By discovering membranes in his solution to the multiple string theories problem, Edward Witten unwittingly developed a framework to solve all these problems as will soon be illustrated.[1]

Exactly how would a collision of membranes transfer energy into a new universe? Current theory is not entirely clear on this subject, and the result is a big bang starting from a singularity, which has many of its own mathematical difficulties. The beauty of string theory is that it does away with the need for any singularities because the smallest anything can be is the Planck length of 10^{-33} cm (10^{-35} m), as that is the size of a single string.[2] For example, the core of a black hole may no longer be a singularity but just extremely small instead. A neutron star packs the mass of an enormous star into a sphere just a few miles across because gravity overcomes the repulsive forces separating the protons from the electrons in the star's atoms once fusion in the star ends, and it collapses. Compared with a neutron star, a black hole has so much mass that it could overcome even the nuclear forces that hold the three quarks (that make up each neutron in the neutron star) apart from each other. This would mean that the center of each quark (according to M-theory, a single string) would no longer have the strength to overcome the force of gravity, and the field that makes up the quark's shell (similar to electrons orbiting the nucleus of an atom) would collapse. This would cause each quark's central string to become compressed almost directly against its neighboring strings, preventing those strings from vibrating in all dimensions, slowing time nearly to a complete stop for those strings within the black hole.

Therefore, a Type I black hole (the mass of about twenty suns to more than thirty-five suns) could have a core just several tens of meters across made of highly compressed individual strings. This would curve spacetime so much that it would create an event horizon far away from the tiny core of strings inside the black hole. The compressed bundle of strings at the core would be so tightly compacted that they could no longer vibrate independently of each other. This would cause the strings at the core to act as if they were a single elementary particle because each black hole is left with only three defining properties: mass, force charge, and spin, similar to a single electron.[3] The center of a black hole would no longer be a one-dimensional "singularity" (or a tear in spacetime) as was originally theorized. The core of the black hole is merely so much smaller than the enormous gravitational forces it produces. The event horizon of the black hole would continue to grow even farther away from its core as the black hole swallows more matter (because the core's size grows so little relative to the gravitational effects produced by the matter falling into the hole). I developed this hypothesis in 2003, and in 2004 other string theorists developed theories along the same lines—that information is not destroyed within a black hole, as it would be within a singularity, but is recycled as Hawking radiation with the same information as the matter that was swallowed by the hole.[4] During the last two years, most physicists have come around to this point of view; that information is recycled from black holes as Hawking radiation, eliminating the singularity.

Singularities are likely not the answer for either black holes or the beginning of the universe, because they also create nonsensical mathematics. After considering how branes collide and vibrate, it appeared there had to be a direct method of energy transfer from the membrane to the new universe at the fundamental level of each string. Strings and branes interacting with each other seems possible as they are both made of the same material, with one being stretched out to an enormous degree compared to the other. Strings may simply be two-dimensional tubes of our brane that were separated from it by quantum noise creating virtual strings. If energy were directly transferred from a brane-brane collision to the individual (virtual) strings within them at the instant of impact, it would mean that matter's mass is not tied up in loops of string which have no interaction with our brane, as M-theory currently states. Rather, *mass is the effect that a string's gravitational vibrations have upon the membrane within which it travels*. Imagine a string vibrating within a four-brane as represented in fig. 1. If a string has the type of vibrations to act as matter, the string's vibrations in the gravity dimensions will curve the membrane around it within a four-brane) as it travels, which curves spacetime (as is consistent with relativity). Therefore, our membrane now *is* spacetime. In the original version of string theory (before M-theory) according to Brian Greene, the higher the frequency of a string, the more massive the particle it emulates.[5] This is also true within the *membrane theory of gravity*: the higher the frequency of a string, the more energy it has, and the further its vibrations will curve the membrane around it, creating the force of gravity.

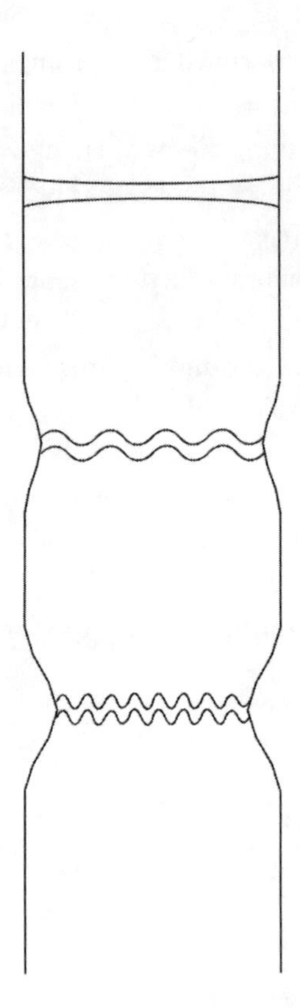

Fig. 1. *A representation of an "MRI view" of a membrane, strings, and their curvature of spacetime. In this sliced-open diagram, the lines on the left and right represent our membrane and its curvature. The three horizontal strings (hollow two-dimensional tubes) are (top) a virtual string that hardly vibrates at all and is basically a hollow tube that was separated from our brane by quantum noise; (middle) a string that represents a quark or other massive particle given energy from the big splash; and (bottom) a string that represents a more massive strange, charmed, or similar quark given additional energy from a high-velocity impact with another string whose mass (vibrations) curves our brane even farther, creating gravity. The more periods the string has (that is, the more rapidly it vibrates), the more it contracts the membrane due to its vibrations and motion and the more massive the particle it emulates.*

In this modification of M-theory, it may be simple string geometry and dynamics that create the effects of gravity. When a virtual string is disturbed into existence by quantum noise, it is a two-dimensional tube of our brane that is separated from our brane in its center but still connected to the brane on either end. While still virtual, the string has incredibly slight vibrations that have very little effect on our membrane. When struck by a big splash, however, a virtual string is given enough energy to vibrate rapidly and become a string of matter, such as an electron or a quark. As the string vibrates, the distance between either end of the string should decrease as more of the string's length is pushed farther away from the plane of the string's vibrations (see fig. 1). The stronger the vibrations of each string, the more that its ends will be pulled together, curving our brane and creating the "force" of gravity. The more energetic a string's vibrations, the heavier the particle it emulates. Gravity is a side effect of the vibrations of strings within our membrane. When fusion occurs within a star, four hydrogen atoms are fused into a single helium atom, and some of the hydrogen's mass is lost in the process, having been converted into pure energy. When this occurs, one of the rapidly vibrating strings loses some energy by emitting a photon, and that string will vibrate less frequently, like switching from the bottom string diagram to the middle string diagram in fig. 1. This pushes the brane apart slightly, sending out a brane vibration (called a graviton) that has only the gravitational force and no other forces associated with it.

This means that, if we run the expansion of the universe backward from its current point to the big bang itself, the previous theorists simply went too far back and assumed that the universe started from a single one-dimensional point before it inflated. Instead of the singularity of the standard model of the big bang, a *membrane theory of gravity* states that when our brane collided with another one, the vast majority of that energy was transferred directly to each membrane in the form of vibrations, which, in this theory, become dark matter. Additionally, at the moment of impact of the big splash, approximately 4 percent of the brane would be in the process of generating virtual strings. This creates an outlet other than the brane itself where the energy of the collision could go, transferring those vibrations directly to the virtual strings within the brane, creating all matter and antimatter in the big splash. According to Brian Greene, Michio Kaku, and a number of other physicists, even "nothing" is highly unstable on the quantum level. Due to quantum noise (see fig. 2) on microscopic scales (scales at or below the Planck length of 10^{-33} cm) tiny bubbles can start to form. These bubbles (strings) vibrate in eleven dimensions (one of which is time), and when they are given energy from an external source (such as a brane collision), they can vibrate in ways that make them take on the properties of all matter and energy. When virtual strings are generated, a quantum disturbance generates tiny waves of spacetime (our membrane), so that one could observe a pair of virtual strings orbiting each other briefly before they merge back together and vanish as their waves cancel each other out.

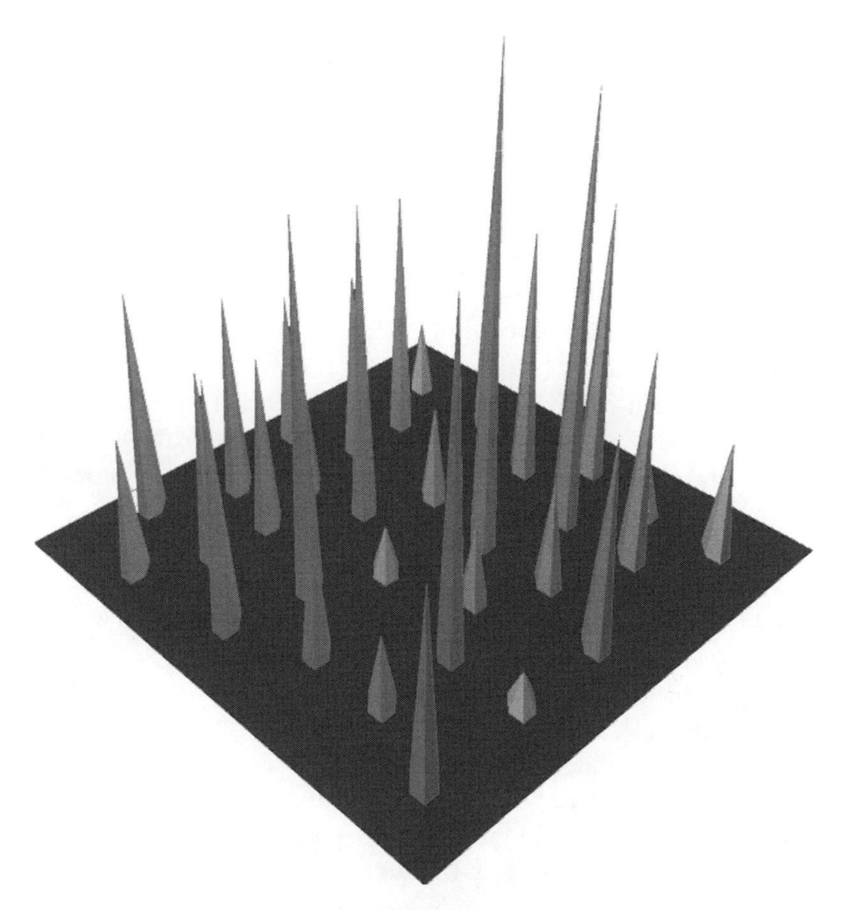

Fig. 2. *Computer models of quantum noise. On Planck-length and smaller scales, this noise appears as tiny bouncing spikes of static, similar to a randomized graphic equalizer on a stereo bouncing to radio static. This noise agitates spacetime on tiny Planck-length scales, forming tiny vibrating strings of spacetime. These strings can inflate into huge membranes, from which other virtual strings can then be separated whose ends are connected to the brane. Eventually, huge 4-branes (with 3 spatial and one time dimensions) full of virtual strings generated from this quantum noise gain momentum as more and more strings are created and destroyed within them and interact briefly with each other. With all this activity, it is likely a matter of time before two of these large membranes will collide with each other, creating new universes.*

This explains gravity in a completely different manner than both string and M-theories currently do. M-theory states that gravity is a closed-loop string that can escape our membrane, accounting for gravity's weakness. However, the result of this is that the only way to satisfactorily explain how gravity works is by adding multiple universes and membranes from which gravity can leak to our membrane. Lisa Randall has postulated such a mechanism on the BBC TV series *Horizon*.[6] The chances that two randomly generated membranes would have this kind of symbiotic relationship are highly unlikely, because branes are only created so often and two of them would need to be created at exactly the same time and place, one within the other. The much more likely scenario is that membranes are independent and self-contained structures that were randomly generated from the quantum noise that exists everywhere in spacetime. Not a complex and highly unlikely arrangement of interconnected parallel universes as M-theory is beginning to require with its vanishing version of gravity. By eliminating the closed-loop strings of M-theory and replacing them with the simple and straightforward curvature of our membrane caused by the vibrations of strings of matter, we eliminate the necessity of relying on a contortion of parallel universes to explain the laws of nature. The universe is an elegant creation, even if its mathematics are turning out to be more complex than we may like. Simplicity requires we recognize that the most likely solution is *membrane gravity*, not an unlikely relationship between randomly generated branes.

If this *membrane theory of gravity* is correct, then what came before our universe could have been these tiny multidimensional strings or bubbles of energy being generated from the quantum background noise, such as Kaku has postulated on *Horizon, the* BBC TV series. Similar to how a breeze can cause soap film to generate bubbles, and bubbles within bubbles, these vibrating virtual strings (and membranes) could be generated and in the process smooth out spacetime from what were rough, choppy spikes of quantum noise. If, over billions or trillions of years, more membranes and virtual strings within them bubble into existence because of this instability, these strings and branes could then stretch, inflate, and interact with each other, creating a much more complex and dynamic system. Eventually, these membranes and the virtual strings being created and destroyed within them could fill up huge volumes of space that might be billions, trillions, or more times larger than our own universe.

Some of the first membranes to evolve could expand into huge sheets, bubbles, and into many other multidimensional shapes like a torus (the shape of a doughnut) when they inflate from the vacuum around them, which has negative pressure (like suction).[7] Then, virtual strings could be disturbed into existence within these branes, possibly inflating them further. Virtual strings generated within these branes have the potential to become matter and energy if they are vibrated rapidly enough in a collision of branes or "big splash."

The initial motion of the brane combined with the generation and interactions of virtual strings within each brane give them momentum as they expand. As a brane's virtual strings interact with each other, the brane and its surface could start to bend and ripple similar to waves on an ocean as other string theorists have postulated, leading to their theory of a membrane big splash. With all this motion initiated by quantum noise, it is only a matter of time before two nearby branes could collide. Eventually, a multitude of branes would be generated and travel through the eleventh dimension, colliding with greater force and frequency as their numbers increase. However, M-theory's current big splash model lacks a clear mechanism of energy transfer from the collision to the new universe. Their result is similar to the big bang theory where initial conditions determine the physics of the new universe, unlike the *membrane theory of gravity* where physics stays intact throughout the collision.

This new brane theory could also explain why normal matter is such a small percentage of the mass of the universe. This could be because only about 4 percent of spacetime within our brane was in the process of forming virtual strings at the instant of the brane-brane collision.[8] These virtual strings created an outlet other than the brane itself where the energy of the collision could go, vibrating the virtual strings and creating matter. Any energy not transferred directly to the virtual strings within our brane would have simply vibrated the membrane itself, creating the vast majority of mass (over 90 percent) as dark matter or brane vibrations (simply the curvature of spacetime). If some strings were left over from previous collisions of this or other branes in addition to the virtual strings, those strings would also be available to be vibrated further into matter and antimatter strings during the collision.

The small percentage of energy that got through to the virtual strings within each brane vibrated the three gravity dimensions of each string. This *directly* transferred energy to the nonenergetic virtual strings within each membrane, creating matter and antimatter quarks and electrons in up and down pairs as will be described shortly. The major brane vibrations that act as dark matter vibrate over large areas thousands of light-years across and even form dark matter galaxies. Huge chains of galaxy clusters are created, attracted to these long strands of dark matter, and form an enormous lattice of strands following the brane vibrations and appearing similar to a three-dimensional spider's web. This *membrane theory of gravity* would help to explain these huge structures as well, because as our brane was compressed during the collision due to its vibrations, a network of crossing membrane waves would form, similar to a crystal lattice. These vibrations of the membrane itself would cause it to contract quickly into a much smaller volume as the membrane vibrates rapidly back and forth. It would form huge waves, both on the outer surface and throughout the inner structure of a four dimensional membrane (see fig. 3). The dark matter resulting from brane vibrations could also cause the gases of the early universe to collapse and form into stars and galaxies earlier than one billion years after the big splash, much more rapidly than current theories including the standard model and M-theory can explain.

Fig. 3. *A higher-dimensional view of two four-branes (with three spatial dimensions and one time dimension) colliding. As they collide, they create "big splashes" on each membrane radiating from the point of the collision. Both branes contract the most at the point of impact due to the violent vibrations. The vibrations would similarly reverberate throughout the center of our spherical four-dimensional brane, creating dense vibrating ripples of membrane (dark matter) throughout the new universe. The vibrations form a network of dark matter which helps to accelerate the formation of the new universe as the young gases are attracted to these dense vibrating dark matter ripples. Additionally, this collision starts off our universe at a size much larger than previously theorized, doing away with the need for any singularities. Therefore, physics never breaks down in this model of our universe, unlike the big bang model where all of physics breaks down before* 10^{-43} *seconds into the expansion from the so-called singularity. The universe simply contracts from the impact, and then starts to expand from a point possibly billions of light-years across. Even without the larger starting volume, our current minimum diameter is at least 156 billion light-years across, according to the WMAP science team. These branes would not appear smooth before the collision as they do in this illustration (for simplification). Instead, they would ripple slightly due to quantum noise, and those ripples would be imprinted on our universe. They appear as areas of higher and lower energy impacts where the ripples constructively and destructively interfere with each other at the instant of the big splash. This pattern is visible in the cosmic background radiation left over from the splash as we will see in fig. 5.*

The vibrations of our membrane are not contained within a small, confined space as the vibrations of regular strings are. (The strings vibrating within our universe that act as matter and energy and are confined to the Planck length). Because brane vibrations have our entire membrane to expand and vibrate throughout, the vibrations are not stable and will dissipate throughout our entire brane. They radiate according to the inverse square law as the waves expand in three dimensions and dissipate throughout our universe into the rest of our brane (accounting for gravity's weakness). As our universe expands, the dark matter brane vibrations will slowly be converted into dark energy as those vibrations relax and dissipate, pushing points that were close together farther apart (see fig. 4). This will happen as the vibrations of our membrane occur less frequently, especially in the empty (starless) regions of the universe between galaxy cluster strands. The cooling of our universe will start to convert the potential energy of our vibrating membrane (dark matter) into kinetic energy (dark energy). This release of potential stored vibrating brane energy pushes our universe apart on an accelerating curve, starting slowly, then increasing in speed until it relaxes closer to its original size before the splash at which point the acceleration will decrease and eventually end as shown in fig. 4. Additionally, during the early universe, the large number of massive gravitational waves produced from orbiting black holes, supernovae, hypernovae, and other gravity wave sources may have aided in sustaining our membrane's rapid vibrations.

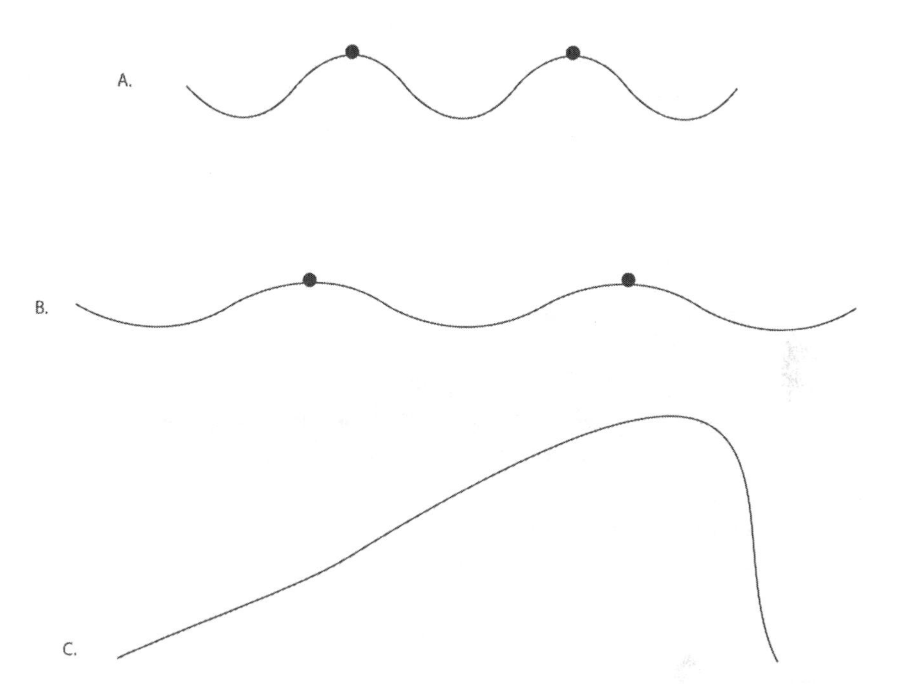

Fig. 4. *Vibrations relaxing and the acceleration of our universe's expansion due to membrane gravity (not to scale). As our membrane relaxes, two points that were closer together (A) get pushed farther apart (B) as the spacetime between them vibrates less frequently and becomes less compressed. As this occurs throughout our universe, it expands more rapidly until the brane relaxes almost completely, at which point the expansion slows down. Because of the rate at which brane vibrations relax, spreading out spacetime, dark energy (C) the rate of acceleration of our universe should appear fairly constant in the beginning. The universe's acceleration begins slowly after the big splash due to the contraction of our brane, then its rapid vibrations from the collision. It continues to accelerate in the middle as dark matter is being constantly converted intodark energy, and will eventually slow down as the brane expands to its original, relaxed form, which can expand no further. No other theory predicts that dark energy will slow down then stop as our universe reaches its original size as this membrane theory of gravity does.*

After about eight billion years of our universe's expansion after the splash, it cooled down significantly, and many of the gravity wave sources decreased at the same time our universe expanded and diluted the original brane vibrations from the big splash. The massive energies given off by quasars stopped as these galaxies containing feeding black holes settled down into normal galaxies like ours. Their central black hole(s) merged and stopped feeding, and new star formation and destruction slowed its pace, reducing the number of supernovae that produced gravity waves as well. As our universe ages, the expanding empty (starless) spheres of space between galaxy clusters strands begin to have very few gravity waves passing through them at the same time our brane relaxes even further. This brane relaxation begins to accelerate the expansion of our universe. Because these regions are isolated from the network of brane vibrations created during the big splash and from any gravity wave sources within faraway galaxies, any vibrating membrane waves in these regions begin to relax and dissipate by this time. Due to the *membrane theory of gravity*, the relaxation of any original brane vibrations in these regions cause their dark matter vibrations to be converted into dark energy expansion forces. The brane relaxes and spreads back out the most in these regions first, as they will elsewhere throughout our brane as its vibrations continue to relax and accelerate our expansion.

This *brane theory of gravity* could also explain other mysteries of our universe: why gravity is so weak and why so little of our universe's mass is made up of normal matter and energy that we can see. Our membrane began life as a single four dimensional string. This brane eventually inflated to a size much larger than our own universe. When our brane was young and small, its interactions (if any) with strings of matter could have been much more powerful, similar to the electromagnetic interactions of strings today, which are 10^{39} times more powerful than the force of gravity.[10]

As our brane grew from the size of a string to a size larger than a galaxy and then larger than many galaxies, its surface area inside and out continued to expand. As it did, the effect of vibrations of standard Planck-length strings of matter (which curve our brane) got continuously weaker as those tiny strings stayed the same size while our brane continued to expand. The gravitational effect that only one of our strings of matter, such as a quark or electron, has on our brane continued to decrease as it had a smaller percentage of the membrane's surface area to vibrate against. When our brane finally reaches its current size, the vibrations of strings of matter have an infinitesimal effect on the tiny percentage of the brane that they vibrate against. Therefore, according to the *membrane the-*

ory of gravity, gravity's weakness should be directly proportional to the size of our membrane.

Because of this, it may be possible to calculate the approximate size of our membrane. We do this by comparing the strength of gravity to the electromagnetic force (as they may have been comparable when our brane and a single string of matter were of equal size) and then by factoring in the distance that the electromagnetic force acts over. The approximate distance between a proton and an electron in a hydrogen atom (the Bohr radius) is about $5 * 10^{-11}$ m. If we multiply this by the difference in power between the two forces (10^{39} times), we get $5 * 10^{30}$ m. A light-year is $9.5 * 10^{15}$ m, so if we divide our result by the distance in a light-year, we get $5.26 * 10^{14}$ light-years or a maximum radius of 526 trillion light-years across (roughly). If our universe is at least 156 billion light-years in diameter as a recent WMAP science team study found[11] (and possibly larger if it began in the manner described earlier), then our membrane could be up to six thousand times larger than our universe currently is. As our universe continues to expand, it has plenty of room to grow, and shouldn't reach it's maximum size for at least a hundred billion years.

This method of universe creation would also explain why only matter and not light was created directly in the big splash. Where virtual strings existed within our brane during the moment of impact of the big splash, all those strings were struck in their three gravity dimensions. Because this impact influenced all three of our string's gravity dimensions at once (as those are the dimensions that interact with our brane), it caused the strings to vibrate in at least all three brane curving dimensions of gravity. This created string vibrations with mass, such as quarks and electrons in matter and antimatter forms, depending on the strength of the impact where each pair of virtual strings existed. The collision could not create photons directly as all three matter dimensions were impacted, and light only vibrates in two of our three gravity dimensions. However, light would soon be created as matter and antimatter strings annihilated each other, creating gamma rays. As soon as our universe went through its final phase change into matter, it behaves exactly as the big bang model does.

A *membrane theory of gravity* could also change our idea of what multiple universes are within other membranes. Current theory states that the initial conditions of a big bang or big splat determine all the physics of the universe created from it. It is currently believed that other universes will all be very different, and that it just happens to be luck that initial conditions were so favorable for life in our universe. However, if *membrane gravity* is correct, all universes created within four-branes like ours will have nearly identical laws of physics. Only the *strength* of gravity will differ, depending on the size of the brane the universe is created within. Thus the smaller the brane, the stronger the force of gravity—and the more powerful the collision that creates the universe, the more energetic the matter and antimatter strings that will be created within it. If a brane collides more gently than ours did, less matter and hardly any antimatter would be created within it. If the brane were smaller than ours as well, its stronger force of gravity could compensate for the smaller collision to create a viable universe, habitable by creatures like us. If, on the other hand, an enormous four-brane has a minor collision with another brane, the small amount of matter created (and its weak gravity) would make for a universe of diffuse gas that may never form a single star. In five-brane, six-brane, and higher spatial dimensional universes, however, the laws of physics would be completely different but the same for each number of dimensions (except for gravity, which would again vary on brane size and the number of spatial dimensions each brane has).

If the *membrane theory of gravity* is correct, it should be *verifiable through observation* using several methods. We could calculate the average mass of space in the universe during the first eight billion years of its existence, mostly in the empty (starless) regions of diffuse gas between galaxy clusters, and compare those results with the average mass of empty space in the universe during the last five billion years. *If our results show that dark matter has been disappearing as our universe's expansion has been accelerating (and losing mass during the last five billion years), it would confirm the membrane theory of gravity and validate string theory simultaneously.* Gravity would not be leaking from our universe as M-theory predicts or originate from a membrane or dimension just slightly away from our own as an updated M-theory of gravity postulates. Another way to confirm the *brane theory of gravity* is by measuring dark energy. *If we find that our universe's acceleration is not precisely constant, but will change and then end over time, (as in Fig. 4), this will also confirm the membrane theory of gravity as no other theory can currently account for this result.* As our membrane's vibrations left over from the big splash relax, any two points on those waves will be pushed apart slowly at first and then more rapidly. Finally, the acceleration will slow down and stop as our brane reaches its original size before the collision.

The energy required to create an electron is much less than the energy required to create a positron, and the same relationship applies to all other types of matter and their antimatter partners. This is fortunate for us, because if matter and antimatter were equally easy to create, then all matter would have been annihilated immediately after the big splash, and the universe would only be made up of cosmic background radiation. If our universe was created by a collision of membranes which transferred kinetic energy directly to each brane, and 4 percent of that energy was transferred directly to the virtual strings within our brane, then the ratio of matter to antimatter during the splash *should be directly proportional to the amount of energy required to create that matter and antimatter.* If the collision imparted its energy directly to the virtual strings, it should do so according to the laws of physics and not break down like the big bang model. After the splash, when matter condenses out, the splash behaves exactly as the big bang model does, with the same ratios of elements created.

This *membrane theory* also shines new light on the cosmic background radiation (CBR). Because the collision of branes struck our universe's strings in our gravity dimensions, it transferred energy creating a vast majority of normal matter in our universe. In those areas of higher energy collisions where brane ripples overlapped, generating constructive interference (and more energy), some higher energy antimatter, such as positrons, was created in those areas. The hot bubble of subatomic strings expanded, and the antimatter was instantly annihilated when it interacted with its normal matter counterpart quarks and electrons, generating gamma rays. The subatomic cloud cooled and condensed into hydrogen and helium atoms. At that instant, a flash of light was emitted. 13.7 billion years later, that light was stretched out as the universe expanded, and we see this light today as the weak microwave cosmic background radiation (see fig. 5).

The Wilkinson Microwave Anisotropy Probe, Image courtesy of NASA, thanks to the WMAP Science Team.[12]

Fig. 5. *NASA's WMAP satellite detected the millionth of a degree difference between the hotter, lighter areas of the cosmic background radiation and the cooler, darker areas. What was once a flash of photons shortly after the birth of our universe—when it condensed into hydrogen and helium gas—has cooled and been stretched over the last 13.7 billion years as our universe has expanded. This has turned light rays into the weak microwave background radiation seen in every direction of the sky, which is the fingerprint of the creation of the universe.*

Because the universe could have been created at such a large size with the *membrane theory of gravity* (possibly billions of light-years across), the quarks and electrons created in the collision may have condensed *almost instantly* (or at least much more rapidly.) This would mean that the cosmic background radiation may not have been created 400,000 years after the big splash as is currently thought. Instead, the CBR could have been created very shortly after the splash and much more rapidly than the "standard model" predicts. Our universe would have rapidly cooled and coalesced from hot quarks and electrons into hydrogen and helium atoms due to the comparatively large size of the early universe.

This would mean that when we look at the CBR, we could be looking *almost directly at an image of the creation of the universe itself.* We may no longer be looking at a picture of our universe nearly 400,000 years later as current cosmologists believe (according to the "standard model" of the big bang). With this new *membrane theory of gravity*, the universe could have started at a much larger size, possibly several billion light-years across, and started expanding from that point. According to the WMAP science team, our universe is at least 156 billion light-years in diameter after 13.7 billion years of expansion and possibly even larger, depending on the starting size of our universe.

Another recent piece of evidence in favor of this *membrane model of gravity* is the Hubble Space Telescope study of 2003 by quantum gravity physicists. They found that the universe is not as "lumpy" as it should be if spacetime were pixilated at the Planck length of 10^{-33}cm (the smallest anything can be according to string theory, as well as the length of a single string). Quantum gravity physicists believe that photons traveling through our universe would move from "pixel to pixel" of Planck length segments as they travel.[13] Images of objects that are farther away should have appeared smeared compared to nearby objects as more distant photons travel through more segments. However, these results were not found. Photos of both nearby and distant objects appeared equally clear. Instead of casting doubt on the Planck length or other solid physics, it is further support for the *membrane theory of gravity* that photons are not traveling through "pixilated" space. Rather they are traveling through a perfectly smooth membrane, which would not blur photons in the least.

We need to move from a particle-based view of the universe to a membrane and string-based view. Researchers are having trouble figuring out dark matter, dark energy, and many other recent discoveries because they view the universe as a collection of particles within a vacuum with "nothing" beyond the edge of our universe. Even M-theory currently suffers from this view as it still does not see our membrane as the source of gravity, with each string currently behaving like a particle in standard physics. The closed-loop graviton string is what is holding M-theory back from becoming a complete theory of physics. The picture of the universe becomes so much clearer when we realize that we live within a fluid-like membrane that is the source of our gravitational force. Because of this, all matter and energy are simply vibrating strings, formed from and connected to our membrane, that travel through our brane and whose vibrations generate all the forces of nature that we observe.

Additionally, inflationary models of the big bang have been under tighter scrutiny in recent years as results from WMAP have been analyzed, and the results do not quite match predictions. This is a very difficult feat, considering there are so many different predictions. The inflationary models have many problems, one of which is quite similar to complaints about the first five string theories before they were unified under M-theory. There are so many different inflationary approaches, with such a wide range of predictions that it has been suggested inflation could never be disproved by observation![14] Here is another problem with inflationary models: What caused the inflation and the "bang" in the first place? This question can never be answered with current models, which is very frustrating to say the least. Some physicists seem not to mind being unknowledgeable of what event actually triggered our universe's creation and expansion. However, it seems vital for any comprehensive theory of our universe to be able to predict how our universe began and how it will end.

In a *Science* magazine article by Adrian Cho, the latest finding by Christoph Adami, of the California Institute of Technology in Pasadena and the Keck Graduate Institute in Claremont, that quantum entanglement is linked to gravity could be one of the strongest pieces of evidence yet for the *membrane theory of gravity*.[15] If two particles such as electrons are entangled, an observer could manipulate one of the particles, and cause it to spin up. (Electrons have two types of spin: one called "up" and the other, "down.") When the particle spins up, they would instantly know that the other entangled particle's spin is down, no matter how far away it is. This interaction makes sense in a *membrane theory of gravity*, because all matter and energy in our universe are connected to each other through our brane. The strings are physically connected to the brane, and their gravitational vibrations curve our brane. Quantum information could only be transmitted or linked if the particles themselves were connected or linked in some way, as they would be through our brane. All matter in the universe—and even light itself, which vibrates in only two gravity dimensions and one higher dimension—is moving through our brane, constantly in contact with it. No other theory can yet account for why quantum entanglement would be linked to gravity, especially current M-theory where gravity is not linked to anything and passes right through our brane as if it were not even there.

Although classical physics can explain light's effects, a membrane model of gravity could make them easier to comprehend. For example, why does light (a massless "particle") exert a tiny force on matter? Solar sails take advantage of this property of light when trillions of photons bounce off a reflective lightweight Mylar sheet and cause acceleration. Because light has no mass, this would seem an impossible task. However, by using M-theory with membrane gravity, we can see that because strings of light vibrate in two of our three gravity dimensions, curving our brane in those two dimensions, light waves can bounce off the brane dimples of the Mylar atoms in those two dimensions. This action transfers a very slight amount of the light string's frequency vibrations to the Mylar as momentum and decreases the light's vibrations (frequency) very slightly from the energy transfer (see fig. 6). This would also explain how light can transfer its energy as heat to other substances. Light causes the molecules to vibrate faster after a collision, because the faster an atom vibrates, the higher its temperature.

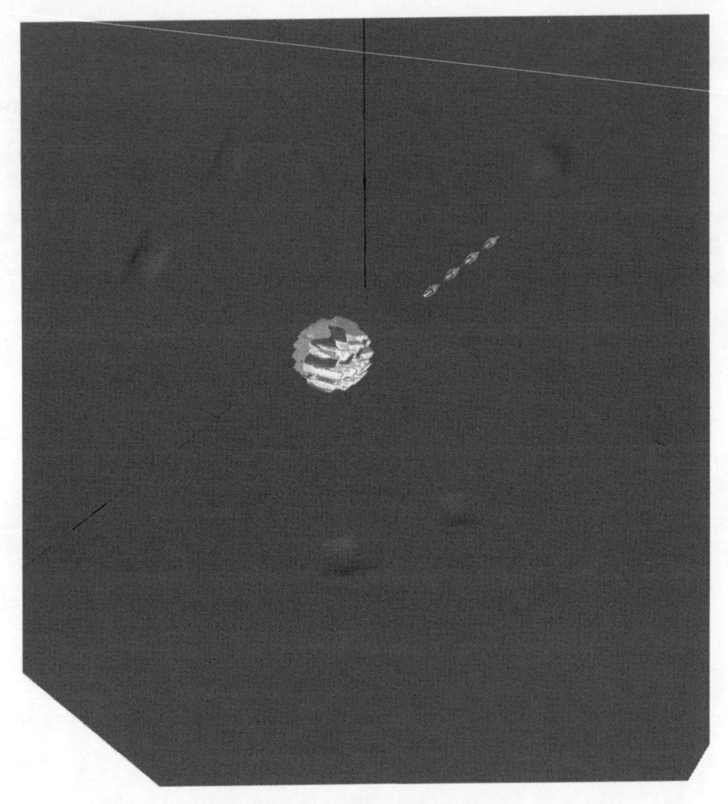

Fig. 6. *Strings of an electron and a photon and their effects on our membrane and each other (exaggerated, not to scale, dimples inverted). The round string on the left represents an electron or quark, which is matter that vibrates in all three of our gravity dimensions. This curves our membrane, creating dimples in all three dimensions (the three dimples pictured below it, to the left, and to the right). The photon of light string (the line with four bulges above and to the right of the string of matter) vibrates in only two of our three gravity dimensions, plus a higher unseen dimension. Therefore, light curves our membrane in only two of its three dimensions (the dimples below it and to the left of it only—the dimple on the right wall belongs to the string of matter, not the photon) and thus has no mass. However, because a photon does curve our brane in two dimensions, it can interact with matter in those two dimensions and transfer vibrations from the photon to the electron through colliding with the electron's brane dimples. Because gravity is so weak, and light only interacts with matter in two of the three gravity dimensions, the energy transfer is extremely weak, taking trillions of photons to have any noticeable effect on a lightweight reflective sheet of Mylar.*

A *membrane theory of gravity* could also explain why $E=MC^2$. As a string of matter vibrates, it curves our membrane in three dimensions. However, when a string of matter is converted into energy, the string will only vibrate in two of the three gravity dimensions, transferring the energy from the one gravity dimension removed into the two remaining gravity dimensions, conserving the string's energy. Therefore, the string's energy remains constant as it vibrates in fewer dimensions. $E=MC^2$ simply illustrates the conservation of energy as the vibrations of a string in three gravity dimensions are compacted into the vibrations of a string in only two dimensions that travel at the speed of light as matter is converted into light.

This theory could also explain why all fundamental strings have two types of spin. If, when the big splash occurred, the only strings available within our brane to transfer energy to were virtual strings that existed at the precise instant of the collision, then the energy of the collision would have been transferred directly to each pair of virtual strings. Each pair of strings would have been given the same amount of energy depending on their location within the brane. One pair, if given enough energy to become electrons, would split apart as their electromagnetic vibrations repelled each other, and one of the electrons would spin up, and the other would spin down. This occurs because the waves of each virtual string in the pair vibrate in the opposite direction of the other, because the pair would cancel each other out when they joined back together and disappeared if they had not been struck by the big splash.

The same reaction would occur with virtual strings given enough energy to become quarks. The pair of virtual strings would be vibrated into matter, forcing each couple of strings apart into up and down quarks (with opposite charges of +2/3 and-1/3, respectively). The pair of quarks would also be "entangled" as they were born from the same virtual pair of strings. After being created, the quarks would join together with two other pairs of quarks to form a proton and neutron (with two up quarks and one down quark and two down quarks and one up quark, respectively, in each).

This method of creation for strings could also explain why electrons (when cooled sufficiently to reduce their electromagnetic repulsion) pair up into up and down sets and flow without resistance through superconductors at very low temperatures. Because up and down electrons were created from opposite wave virtual strings, if they are brought close enough together by cooling them, the electrons will fall toward one another and travel together. This is similar to how gravity's curvature of spacetime causes objects to fall toward one another. Because the electrons pair up, they no longer bounce randomly through the material they travel through as they normally would, heating it up and causing resistance. Instead, they form orderly pairs and flow without resistance and with zero loss of energy, a property we can take advantage of to transfer electricity without loss and to create powerful electromagnets.

It is impossible to tell how many membranes have evolved yet (like the one our universe was created within). The birth of a new universe could be a daily or even hourly event, depending on how many branes have evolved so far. If the *membrane theory of gravity* is correct, we likely live within a multiverse (a set of multiple possible universes) of branes moving through our higher dimensions. This system of strings and branes continues to evolve, getting ever more complex, and building upon itself. Complex chemical reactions made possible from the heavy elements created within the first stars (and dispersed throughout young galaxies when these stars die and explode) will continue increasing the complexity of this system. This will lead to the creation of complex organic molecules. After billions of years of replication, competition, and evolution, these once simple molecules will form beings of such intelligence that they can look back on the universe and actually understand where it came from, how it evolved, and how it will end. The universe really is a thing of beauty.

According to this *membrane theory of gravity*, our universe will end in ice rather than in fire as its expansion is currently accelerating. However, our universe will not continue to accelerate its expansion forever as some astronomers currently believe. As the universe continues to expand and our membrane's vibrations continue to relax, dark matter is being constantly converted into an expansionary energy (dark energy). As our brane continues to expand (back toward its original size before the splash), the more that dark matter (brane vibrations) will evaporate (relax). Our universe will end up being a cold, dark place, but the galaxies still remaining in our ancient universe will cling together until our brane expands back out to its original size, and they lose the dark matter holding them together. At this point, all our galaxies will fly apart, sending their stars out into the darkness. The last of the stars will eventually die out, and long after that the remaining black holes will evaporate, and this universe will soon be ready for the next membrane collision to start the process all over again.

It should also be clear by now that our own universe is finite, not infinite, being contained within a single membrane of which there are a multitude. The multiverse, however, may or may not be infinite. It is big, but do not let that fool you. If there are still strings and branes being created out there today, then the multiverse could be finite as well, just very large and still growing.

These modifications of M-theory avoid the apparent breakdown of physics we encounter during the creation of our universe at times before 10^{-43} seconds after the big bang, when the entire universe was supposedly compressed into a singularity. This new version of the big splash with a *membrane theory of gravity* does away with all the breakdowns of physics, infinite energies, and infinitely small sizes of the old big bang model. This leaves a much more stable and robust theory for the creation of our universe. It also predicts dark matter and dark energy and how they change over time which neither the standard model or M-theory can do. It also explains why gravity curves spacetime and why spacetime acts similar to a fluid (as branes do). It predicts a "flat" universe (as we have observed) and among other recent observations it explains, it illustrates how quantum entanglement is linked to gravity, and it behaves just like the big bang after our universe goes through its final (and only) phase change into matter. Once rigorously tested, this hypothesis could finally give physics a true "Theory of Everything" and successfully combine relativity with quantum mechanics through M-theory and this *membrane theory of gravity*.

Notes

Preface

1. Charles Seife, "Big Bang's New Rival Debuts with a Splash," *Science* 292, no. 5515 (2001): 189–91.

Chapter 1

1. Planck length: This tiny unit of measure is the length of a fundamental string in string theory and M-theory, and it is the length of the higher dimensions in which these strings vibrate, creating the "forces" of nature. If a quark were expanded to the size of our solar system, a string would only be as tall as a single tree.
2. Michio Kaku, *Hyperspace* (New York: Oxford University Press, 1994), 99.
3. *Ibid,* 140.
4. *Ibid,* 16.
5. Brian Greene, *The Elegant Universe* (New York: Vintage Books, 1999), 146.
6. *Ibid,* 148.
7. *Ibid,* 148
8. Rod Nave, HyperPhysics Web site, Georgia State University, 2006.
9. Brian Greene, *The Elegant Universe* (New York: Vintage Books, 1999), 144–45.
10. Michio Kaku, *Hyperspace* (New York: Oxford University Press, 1994), 147.

11. *Ibid*, 147.

Chapter 2

1. Brian Greene, *The Elegant Universe* (New York: Vintage Books, 1999), 314.

Chapter 3

1. The Internet Encyclopedia of Science "Wormhole," David Darling, http://www.daviddarling.info/encyclopedia/W/wormhole.html.
2. Wikipedia, "Alcubierre drive," 2006.
3. Erica Hupp and George Deutsch, NASA's Gravity Probe B Mission Completes Data Collection, 2005.
4. Adrian Cho, "Pulsars' Gyrations Confirm Einstein's Theory," *Science* 313, no. 5793 (2006): 1556-57.
5. Theodore C. Nason, e-mail message to author, March 2005.

Chapter 5

1. Paul Recer, "Theory Links Ancient Extinction to Supernova," Associated Press, January 7, 2004.
2. Arnon Dar, Ari Laor, and Nir Shaviv, "Life Extinctions by Cosmic Ray Jets," *Physical Review Letters (*June 29, 1998).
3. David Perlman, "Mass extinction comes every 62 million years, UC physicists discover," *San Francisco Chronicle*, March 10, 2005.
4. Vic Camp, "Climate Effects of Volcanic Eruptions," San Diego State University, 2006.
5. A.V. Fedorov, et al., "The Pliocene Paradox (Mechanisms for a Permanent El Niño)," *Science* 312, no. 5779 (2006): 1485–89.
6. Michael Hopkin, "Warming seas cause stronger hurricanes," news@nature.com, March 2006.
7. Mark F. Meier and Mark B. Dyurgerov, "How Alaska Affects the World," *Science* 297, no. 5580 (2002): 350–51.

8. Information Please Database, "The Sun," Pearson Education, http://www.infoplease.com/ipa/A0004432.html.

Chapter 7

1. York Dobyns, "Retrocausal Information Flow," Princeton University, Google Video, 2006.
2. Blanka Rogina, Robert A. Reenan, Steven P. Nilsen, and Stephen L. Helfand, "Extended Life-Span Conferred by Cotransporter Gene Mutations in Drosophila," *Science* 290, no. 5499 (2000): 2137–40.
3. Jennifer Couzin, "Boosting Gene Extends Mouse Life Span," *Science* 309, no. 5739 (2005): 1310–11.
4. John Dubinski, The Merger of the Milky Way and Andromeda Galaxies, (2001), http://www.cita.utoronto.ca/<dubinski/tflops/.

Chapter 8

1. Brian Greene, *The Elegant Universe* (New York: Vintage Books, 1999), 335.
2. Charles Seife, "A General Surrenders the Field, But Black Hole Battle Rages On," *Science* 305, no. 5686 (2004): 934–36.
3. Brian Greene, *The Elegant Universe* (New York: Vintage Books, 1999), 127.
4. *Big Thinkers*, "Michio Kaku's Superstring Symphony," Tech TV, February 22, 2002.
5. Mario Livio, "Cosmic Explosions in an Accelerating Universe," *Science* 286, no.5445 (1999): 1689–90.
6. Alan H. Guth and David I. Kaiser, "Inflationary Cosmology: Exploring the Universe from the Smallest to the Largest Scales," *Science* 307, no. 5711 (2005): 890.
7. Brian Greene, *The Elegant Universe* (New York: Vintage Books, 1999), 144–45.
8. *Horizon*, "Parallel Universes," BBC TV, February 14, 2002.
9. The Science Channel, *Top Science Stories of 2004*, 2004.

10. *Horizon,* "The Death Star," BBC TV, October18, 2001.

11. Adrian Cho, "Galaxy Clusters Bear Witness to Universal Speed-Up," *Science* 304, no. 5674 (2004): 1092.

12. Los Alamos National Laboratory, The Electromagnetic Force (2000), http://public.lanl.gov/alp/plasma/EM_forces.html.

13. David Whitehouse, "Astronomers size up the Universe," BBC News Online (2004).

14. The Wilkinson Microwave Anisotropy Probe; Image courtesy of NASA, thanks to the WMAP Science Team. I acknowledge the use of the Legacy Archive for Microwave Background Data Analysis (LAMBDA). Support for LAMBDA is provided by the NASA Office of Space Science. http://lambda.gsfc.nasa.gov/product/map/current/m_images.cfm

15. Alan H Guth and David I. Kaiser, "Inflationary Cosmology: Exploring the Universe from the Smallest to the Largest Scales," *Science* 307, no. 5711 (2005): 886.

16. John Whitfield, "Sharp Images Blur Universal Picture," *Nature,* March 31, 2003.

17. James Glanz, "Which Way to the Big Bang?" *Science* 284, no. 5419 (1999): 1448–51.

18. Adrian Cho, "To Escape From Quantum Weirdness, Put the Pedal to the Metal," *Science* 309, no. 5742 (2005):1801.

Chapter 9

1. Charles Seife, "A General Surrenders the Field, But Black Hole Battle Rages On," *Science* 305, no. 5686 (2004): 934–36.

Appendix

1. Brian Greene, *The Elegant Universe* (New York: Vintage Books, 1999), 314.

2. Alan H. Guth and David I. Kaiser, "Inflationary Cosmology: Exploring the Universe from the Smallest to the Largest Scales," *Science* 307, no. 5711 (2005): 886.

3. Brian Greene, *The Elegant Universe* (New York: Vintage Books, 1999), 335.

4. Charles Seife, "A General Surrenders the Field, But Black Hole Battle Rages On," *Science* 305, no. 5686 (2004): 934–36.

5. Brian Greene, *The Elegant Universe,* 144–145.

6. *Horizon*, "Parallel Universes," BBC TV, February 14, 2002.

7. Mario Livio, "Cosmic Explosions in an Accelerating Universe," *Science* 286, no.5445 (1999): 1690.

8. Alan H. Guth and David I. Kaiser, "Inflationary Cosmology: Exploring the Universe from the Smallest to the Largest Scales," *Science* 307, no. 5711 (2005):890.

9. Adrian Cho, "Galaxy Clusters Bear Witness to Universal Speed-Up," *Science* 304, no. 5674 (2004):1092.

10. Los Alamos National Laboratory, The Electromagnetic Force, 2000.

11. David Whitehouse, "Astronomers size up the Universe," BBC News Online (2004).

12. The Wilkinson Microwave Anisotropy Probe. Image courtesy of NASA, thanks to the WMAP Science Team. I acknowledge the use of the Legacy Archive for Microwave Background Data Analysis (LAMBDA). Support for LAMBDA is provided by the NASA Office of Space Science, http://lambda.gsfc.nasa.gov/product/map/current/m_images.cfm.

13. John Whitfield, "Sharp Images Blur Universal Picture," news@nature.com, March 31, 2003.

14. James Glanz, "Which Way to the Big Bang?" *Science* 284, no. 5419 (1999): 1448–51.

15. Adrian Cho, "To Escape From Quantum Weirdness, Put the Pedal to the Metal," *Science* 309, no. 5742 (2005):1801.

Bibliography

Big Thinkers, "Michio Kaku's Superstring Symphony," Tech TV, February 22, 2002.

Camp, Vic. "Climate effects of Volcanic Eruptions." San Diego State University, http://www.geology.sdsu.edu/how_volcanoes_work/climate_effects.html.

Cho, Adrian. "Galaxy Clusters Bear Witness to Universal Speed-Up." *Science* 304, no. 5674 (2004): 1092.

Cho, Adrian. "Pulsars' Gyrations Confirm Einstein's Theory." *Science* 313, no. 5793 (2006): 1556-57.

Cho, Adrian. "To Escape From Quantum Weirdness, Put the Pedal to the Metal." *Science* 309, no. 5742 (2005): 1801.

Couzin, Jennifer. "Boosting Gene Extends Mouse Life Span." *Science* 309, no. 5739 (2005): 1310–11.

Dar, Arnon, Ari Laor, and Nir Shaviv. "Life Extinctions by Cosmic Ray Jets," *Physical Review Letters* 80, no. 26 (June 29, 1998): 1999.5813–16.

Dubinski, John. "The Merger of the Milky Way and Andromeda Galaxies." University of Toronto, http://www.cita.utoronto.ca/~dubinski/tflops/.

Fedorov, A. V., et al. "The Pliocene Paradox (Mechanisms for a Permanent El Niño)." *Science* 312, no. 5779 (2006): 1485–89.

Glanz, James. "Which Way to the Big Bang?" *Science* 284, no. 5419 (1999): 1448–51.

Greene, Brian R. *The Elegant Universe*. New York: Vintage Books, 1999.

Guth, Alan H., and David I. Kaiser. "Inflationary Cosmology: Exploring the Universe from the Smallest to the Largest Scales." *Science* 307, no. 5711 (2005): 890.

Hopkin, Michael. "Warming seas cause stronger hurricanes." news@nature.com, http://www.nature.com/news/2006/060313/full/060313-12.html .

Horizon, "Parallel Universes," BBC TV, February 14, 2002.

Horizon. "The Death Star." BBC TV, October18, 2001.

Hupp, Erica, and George Deutsch. "NASA'S Gravity Probe B Mission Completes Data Collection." NASA, http://www.nasa.gov/centers/marshall/news/news/releases/2005/05-160.html.

Information Please Database. "The Sun." Pearson Education, http://www.infoplease.com/ipa/A0004432.html .

Kaku, Michio. *Hyperspace*. New York: Oxford University Press, 1994.

Kaku, Michio. *Strings, Conformal Fields, and M-Theory*. New York: Springer-Verlag, 2000.

Livio, Mario. "Cosmic Explosions in an Accelerating Universe." *Science* 286, no. 5445 (1999): 1690.

Los Alamos National Laboratory. "The Electromagnetic Force." http://public.lanl.gov/alp/plasma/EM_forces.html.

Meier, Mark F., and Mark B. Dyurgerov. "How Alaska Affects the World." *Science* 297, no. 5580 (2002): 350–51.

Nave, Carl R. "Speed of Sound." Georgia State University, http://hyperphysics.phy-astr.gsu.edu/hbase/sound/souspe.html.

Oganessian, Y.T., et al. "Experiments on the synthesis of element 115 in the reaction 243Am(48Ca,xn)291–x115." *APS Physical Review C* 69, 021601(R), (February 2004).

Perlman, David. "Mass extinction comes every 62 million years, UC physicists discover." *San Francisco Chronicle*, March 10, 2005, http://www.sfgate.com/cgi-bin/article.cgi?f=/c/a/2005/03/10/MNGFIBN6PO1.DTL.

Recer, Paul. "Theory links ancient extinction to supernova." The Associated Press, http://www.msnbc.msn.com/id/3900550/.

Rogina, Blanka, Robert A. Reenan, Steven P. Nilsen, and Stephen L. Helfand. "Extended Life-Span Conferred by Cotransporter Gene Mutations in Drosophila." *Science* 290, no. 5499 (2000): 2137–40.

Seife, Charles. "Big Bang's New Rival Debuts with a Splash." *Science* 292, no. 5515 (2001): 189–91.

Seife, Charles. "A General Surrenders the Field, But Black Hole Battle Rages On." *Science* 305, No. 5686 (2004): 934–36.

Whitehouse, David. "Astronomers size up the Universe." BBC News Online, May 28, 2004, http://news.bbc.co.uk/2/hi/science/nature/3753115.stm.

Whitfield, John. "Sharp Images Blur Universal Picture." news@nature.com, http://www.nature.com/news/2003/030324/full/030324-13.html.

Wikipedia. "Alcubierre drive." http://en.wikipedia.org/wiki/Alcubierre_drive, 2006.

Darling, David, "Wormhole," The Internet Encyclopedia of Science http://www.daviddarling.info/encyclopedia/W/wormhole.html.

978-0-595-40822-
0-595-40822-2

Made in the USA